Ratio, fractions, decimals and percentages

Starting fractions 1

A1 (a) Two halves of the same thing are the same size.

(b) Two halves of the group make the whole group. So there cannot be three halves.

(c) A flag which is half black and half yellow cannot have any red.

(d) Four *quarters* make a whole.

(e) Can half a student be away with flu? Half of 25 is $12\frac{1}{2}$!

A2 (a) Your piece is bigger than mine!

(b) Divide your group into three as equally as you can.

(c)

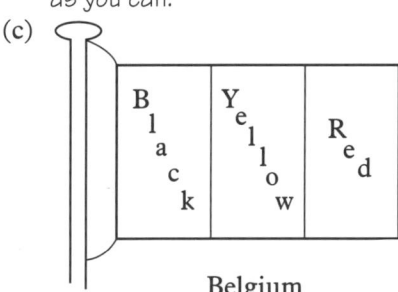

Belgium

The flag of Belgium has three equal parts of black, yellow and red.

(d) Four friends shared the money they collected. Each got a quarter.

(e) About half the students in the class of 25 were away with flu.

A3 Your own sentences.

A4 There are many ways of dividing the square into halves.
Here are four:

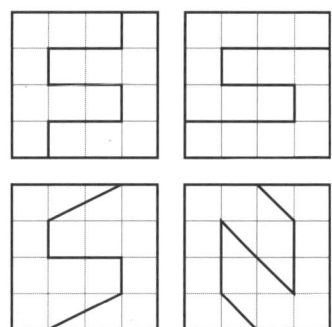

How many ways did you find?

B1 No. The red area is larger than the white area.

B2

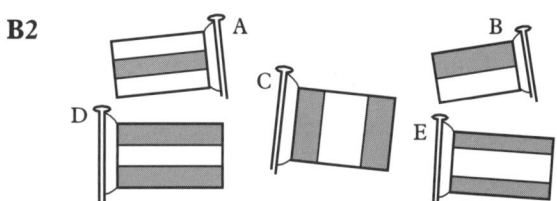

The Austrian flag is D.
Discuss why C and E cannot be the Austrian flag.

B3

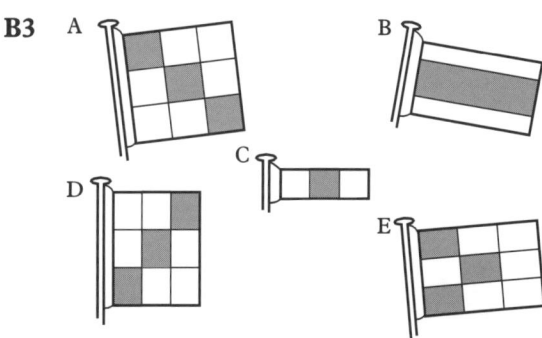

The following flags are coloured one-third red.
A   (There are 9 equal squares and 3 are red.)
C   (There are 3 equal squares and 1 is red.)
D, E   (There are 9 equal rectangles and 3 are red.)
What is wrong with B?

1

B4 ▲ Your own designs.

**Option**

Hungary, Italy and Madagascar all have a flag that is $\frac{1}{3}$ green, $\frac{1}{3}$ white and $\frac{1}{3}$ red. Mauritius has a flag that is divided into quarters. The colours are red, blue, yellow and green.

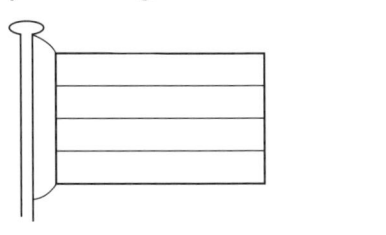

C1 Your own flag. It doesn't have to be exactly like the one in the textbook. It could, for example, look like one of these:

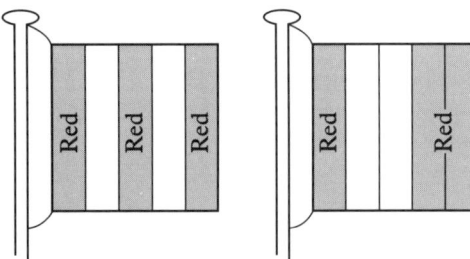

C2 Here is one example:

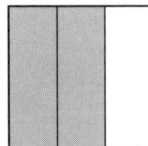

C3 (a) Your own drawing.
(b) The square is split into 6 equal parts.
(c) $\frac{1}{6}$ of the square is coloured.
(d) $1 - \frac{1}{6}$ or $\frac{5}{6}$ of the square is white.

C4 Here are three ways of showing that $\frac{5}{12}$ is greater than $\frac{1}{4}$.

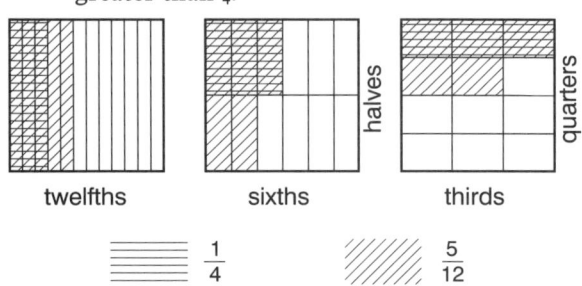

twelfths   sixths   thirds

▤ $\frac{1}{4}$ ▨ $\frac{5}{12}$

C5 You've already used the stencil to split a square into 2, 3, 5, 6 and 12 equal parts. If you draw carefully, the stencil can be used to split a square into the following numbers of equal parts:
4, 8, 10, 15, 16, 18, 20, 24, 30, 32, 36, 40, 48, 50, 60, 64, 72, 80, 96, 100, 120, 144.

Challenge

For good results, you need to fold and crease with great care.
You should have no difficulty in folding a rectangular sheet of paper into 2, 4, 8 or 16 equal parts.
It is possible to fold a sheet of paper into 3, 6, 9 or 18 equal parts without using guesswork. Look at a book on paper folding and see if you can find out how to do this.

C6 ▲

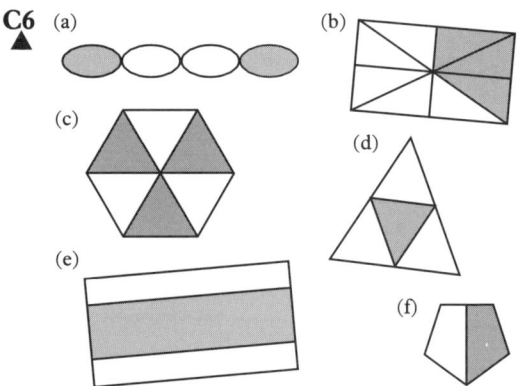

The shapes have these fractions coloured:
(a) $\frac{2}{4}$ or $\frac{1}{2}$ (b) $\frac{3}{8}$ (c) $\frac{3}{6}$ or $\frac{1}{2}$
(d) $\frac{1}{4}$ (e) $\frac{1}{2}$ (f) $\frac{1}{2}$

Before you put your display on the wall, discuss it with some friends to see if they find it convincing.

Starting fractions 2

$\frac{1}{2}$ of 18 is 9.
$\frac{1}{3}$ of 18 is 6.
$\frac{1}{6}$ of 18 is 3.

You can't split 18 counters into 4 or 5 equal piles.

A1 Each person will pay £4.

A2

  ÷ 2 =

You could also press these keys:

0 · 5 × 5 0 =

A3 How many students are in your class? Is it an odd or an even number?

A4 (a) One-third of twenty-four is 8.
(b) Two-thirds of twenty-four is 16.

A5 (a)

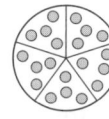 One-fifth of twenty is 4.

(b)

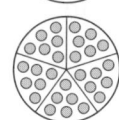 Two-fifths of thirty is 12.

(c)

Four-fifths of ten is 8.

A6 ▲ A   B

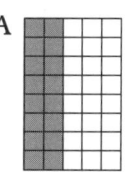 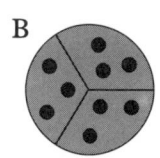

C

D Half of six is three.

E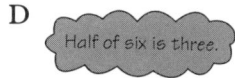

F One-fifth of thirty is six.   G

H
    I

		Answer	Diagram
(a)	Find two-fifths of forty.	16	A
(b)	$\frac{2}{3}$ of 9	6	B
(c)	$\frac{3}{4}$ of £20	£15	E
(d)	$\frac{4}{5}$ of 30	24	F
(e)	Find a quarter of six.	$1\frac{1}{2}$	D
(f)	One-third of £6	£2	G
(g)	$\frac{2}{5}$ of £20	£8	C
(h)	$\frac{3}{4}$ of 12	9	I

A7 $\frac{1}{10}$ of £100 = £10 $\frac{3}{5}$ of £15 = £9
I would rather have $\frac{1}{10}$ of £100 as it is £1 more.

A8 Your own display.

B1 ▲ (a) THREE
(b) DICE
(c) QUARTER

B2 Your own code puzzles.

Investigate

There are many ways of making 1.
Here are four which use different grids:

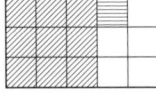

$\frac{1}{3} + \frac{1}{6} + \frac{1}{2} = 1$

$\frac{3}{8} + \frac{3}{8} + \frac{1}{4} = 1$

$\frac{3}{5} + \frac{1}{15} + \frac{1}{3} = 1$

$\frac{1}{5} + \frac{1}{5} + \frac{1}{10} + \frac{1}{2} = 1$

Take it away — if you can!

Is it a good idea to start with exactly 20 counters?
What would you hope to throw if you started with 23 counters in the centre pile?

Fractions 3

A1   (a) $\frac{1}{2}$ of an hour   **30 minutes**

(b) $\frac{1}{4}$ of an hour   **15 minutes**

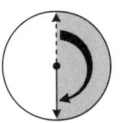

(c) $\frac{3}{4}$ of an hour   **45 minutes**

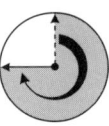

A2   (a) $\frac{1}{3}$ of an hour   **20 minutes**

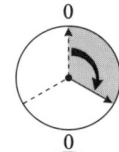

(b) $\frac{2}{3}$ of an hour   **40 minutes**

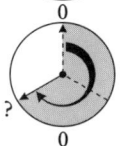

(c) $\frac{1}{5}$ of an hour   **12 minutes**

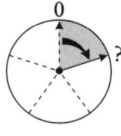

A3   (a) $1\frac{3}{4}$ hours   (b) 6:45 p.m.   (c) 11:45 a.m.
(d) 6:15 p.m.   (e) 6:45 p.m.

A4   (a) 1 hour   (b) $\frac{3}{4}$ hour   (c) 2 hours
(d) $1\frac{3}{4}$ hours   (e) $2\frac{1}{4}$ hours   (f) $1\frac{1}{4}$ hours
(g) $2\frac{3}{4}$ hours   (h) $4\frac{1}{4}$ hours   (i) $3\frac{1}{4}$ hours
(j) 4 hours   (k) $4\frac{1}{2}$ hours

A5   (a) 11:45 a.m.   (b) 9:15 a.m.

A6

Starting time	Cooking time	Finishing time
8:00 a.m.	$2\frac{1}{2}$ hours	10:30 a.m.
11:00 a.m.	$1\frac{3}{4}$ hours	12:45 p.m.
4:30 p.m.	$2\frac{1}{4}$ hours	6:45 p.m.
2:15 p.m.	3 hours	5:15 p.m.
5:10 p.m.	3h 20 min.	8:30 p.m.

A7

Cooking time.
20 minutes for each lb.
plus an extra 25 minutes.

Frozen Chicken 4 lb.

The chicken will take 1 hour 45 minutes to cook.

A8   It takes $1\frac{3}{4}$ hours to travel from London to Leicester.

A9

	Journey	Time
(a)	London to Manchester	4 hours
(b)	Birmingham to Liverpool	$2\frac{1}{4}$ hours
(c)	Leicester to Leeds	$2\frac{1}{4}$ hours
(d)	London to Leeds	4 hours

A10

	Destination	Arrival time
(a)	Leicester	9:45 a.m.
(b)	Nottingham	10:30 a.m.
(c)	Sheffield	11:15 a.m.
(d)	Leeds	12:00 noon

A11 (a) 1:30 p.m.   (b) 2:00 p.m.

B1 (a) $\frac{2}{3}$ of the left-hand strip is coloured.
   (b) $\frac{4}{5}$ of the right-hand strip is coloured.
   (c) $\frac{4}{5}$ is greater than $\frac{2}{3}$.

B2 $\frac{2}{5}$ is greater.

B3 $\frac{2}{3}$ is greater

B4 $\frac{3}{4}$ is greater.

B5 $\frac{3}{4}$ is greater.
▲
B6 (b) is true.

B7 (b) is true.

B8 $\frac{4}{6}$ is equal to $\frac{2}{3}$.
There are many other answers.

B9

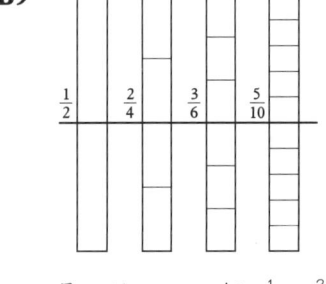

Fractions equal to $\frac{1}{2}$: $\frac{2}{4}$ $\frac{3}{6}$ $\frac{5}{10}$

B10 $\frac{4}{8}$ **B11** $\frac{6}{12}$ **B12** $\frac{8}{12}$

C1 $\frac{1}{3}$ of £12 is £4, $\frac{1}{2}$ of £6 is £3.
I would prefer to have $\frac{1}{3}$ of £12 as it is £1 more.

C2 (a) $\frac{1}{3}$ of £15 is £5.
   (b) $\frac{1}{4}$ of £24 is £6.
   (c) $\frac{1}{4}$ of £24 is more.

C3 $\frac{1}{3}$ of £24 is more than $\frac{1}{4}$ of £24.

C4 $\frac{1}{2}$ of £40 is more than $\frac{3}{4}$ of £20.

D1 $\frac{2}{3}$ is between 0·6 and 0·7.

D2 (a) $\frac{1}{4}$ is between 0·2 and 0·3.
   (b) $\frac{1}{2}$ is equal to 0·5.
   (c) $\frac{3}{4}$ is between 0·7 and 0·8.

D3 (a) $\frac{1}{5} = 0\cdot2$ (b) $\frac{2}{5} = 0\cdot4$
   (c) $\frac{3}{5} = 0\cdot6$ (d) $\frac{4}{5} = 0\cdot8$

D4 $\frac{1}{7}$ is between 0·1 and 0·2.
$\frac{2}{7}$ is between 0·2 and 0·3.
$\frac{3}{7}$ is between 0·4 and 0·5.
$\frac{4}{7}$ is between 0·5 and 0·6.
$\frac{5}{7}$ is between 0·7 and 0·8.
$\frac{6}{7}$ is between 0·8 and 0·9.

D5 $\frac{3}{4}, \frac{5}{7}, \frac{6}{8}$ and $\frac{9}{12}$ lie between 0·7 and 0·8.

D6 $\frac{1}{4}, \frac{2}{7}, \frac{2}{8}$ and $\frac{3}{12}$ lie between 0·2 and 0·3.

D7 (a) The bonnet of the BMW is 0·5 of the length of the body.
   (b) The bonnet of the Farina is 0·4 of the length of the body.
   (c) The bonnet of the Gurney Nutting is 0·3 of the length of the body.
   (d) The saloon of the Silver Ghost is 0·6 of the length of the body.
   (e) The saloon of the Thrupp and Maberly is 0·2 of the length of the body.

Fractions 4

A1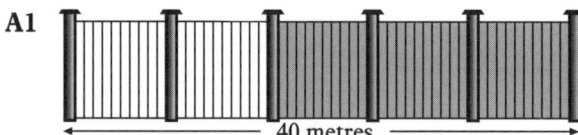

40 metres

Ignore the width of the posts!
(a) $\frac{2}{3}$ of the fence panels are painted.
(b) The painted part is approximately 24 metres long.

A2 You can see $\frac{1}{4}$ of the window.

A3 There are 32 rectangles in the window.

A4 ▲ $\frac{1}{3}$ of £138 = £46,
so $\frac{2}{3}$ of £138 = 2 × £46 = £92.

$2 × £138 = £276$,
so $\frac{2}{3}$ of £138 = £276 ÷ 3 = £92.

You may have found some other ways.

A5 $\frac{6}{7}$ of an iceberg is under the water.
$(1 - \frac{1}{7} = \frac{6}{7})$

A6 (a) 50 cm of the post is below ground.
(b) $\frac{3}{4}$ of the post is above ground.

A7 The two fractions are $\frac{1}{4}$ and $\frac{3}{4}$.

A8 ▲

 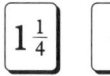

| $1\frac{1}{4}$ | $\frac{1}{4}$ | $\frac{3}{4}$ | $2\frac{1}{2}$ |

| $3\frac{3}{4}$ | $1\frac{1}{2}$ | $\frac{1}{2}$ | $1\frac{1}{4}$ | $\frac{3}{4}$ |

(a) The numbers on the cards add to $12\frac{1}{2}$.
(b) There are nine ways to get a total of 4:
$3\frac{3}{4} + \frac{1}{4} = 4$
$2\frac{1}{2} + 1\frac{1}{2} = 4$
$2\frac{1}{2} + 1\frac{1}{4} + \frac{1}{4} = 4$
$2\frac{1}{2} + \frac{3}{4} + \frac{3}{4} = 4$
$2\frac{1}{2} + \frac{3}{4} + \frac{1}{2} + \frac{1}{4} = 4$
$1\frac{1}{2} + 1\frac{1}{4} + 1\frac{1}{4} = 4$
$1\frac{1}{2} + 1\frac{1}{4} + \frac{3}{4} + \frac{1}{2} = 4$
$1\frac{1}{4} + 1\frac{1}{4} + \frac{3}{4} + \frac{3}{4} = 4$
$1\frac{1}{4} + 1\frac{1}{4} + \frac{3}{4} + \frac{1}{2} + \frac{1}{4} = 4$
(c) (i) $1\frac{1}{2} + \frac{1}{2} = 2$
$1\frac{1}{4} + \frac{3}{4} = 2$

(ii) $1\frac{1}{2} + 1\frac{1}{4} + \frac{1}{4} = 3$
$1\frac{1}{2} + \frac{3}{4} + \frac{3}{4} = 3$
$1\frac{1}{4} + 1\frac{1}{4} + \frac{1}{2} = 3$
(iii) $2\frac{1}{2} + \frac{3}{4} + \frac{1}{2} + \frac{1}{4} = 4$
$1\frac{1}{2} + 1\frac{1}{4} + \frac{3}{4} + \frac{1}{2} = 4$
$1\frac{1}{4} + 1\frac{1}{4} + \frac{3}{4} + \frac{3}{4} = 4$
(iv) $1\frac{1}{2} + 1\frac{1}{4} + 1\frac{1}{4} + \frac{3}{4} + \frac{1}{4} = 5$

B1 The ruler measures halves, thirds, quarters, fifths, sixths, eighths, tenths, twelfths and sixteenths.

B2 $\frac{1}{16}$ is usually the smallest fraction you can measure with a ruler, but some rulers do have $\frac{1}{32}$ divisions marked.

B3 There are about $2\frac{1}{2}$ centimetres in an inch. (An inch is actually a little more than $2\frac{1}{2}$ centimetres.)

B4 (a) $1\frac{15}{16}$ inches
(b) $1\frac{8}{16}$ inches ($= 1\frac{1}{2}$ inches)
(c) $3\frac{2}{16}$ inches ($= 3\frac{1}{8}$ inches)
(d) $4\frac{3}{16}$ inches or $4\frac{4}{16}$ inches ($= 4\frac{1}{4}$ inches)
Even manufactured rulers are not always as accurate as they should be. Your answers might differ slightly from these.

B5 One centimetre is about $\frac{4}{10}$ or $\frac{2}{5}$ of an inch.

B6 A reasonable estimate, using the answer to **B3**, is $12 × 2\frac{1}{2}$ or 30 centimetres in a foot. (A foot is actually almost $30\frac{1}{2}$ centimetres.)

B7 Taking a metre to be about 39 inches, the imperial mile is about 25 metres longer than the metric mile.

B8 $\frac{1}{2}$ $\frac{2}{4}$ $\frac{3}{6}$ $\frac{4}{8}$ $\frac{5}{10}$ $\frac{6}{12}$ $\frac{8}{16}$

B9 ▲ Your own explanation.

B10 Your own lines.
*Try drawing lines of other lengths.
Do you find it easier to estimate in inches or in centimetres?*

Your own report.

6

B11

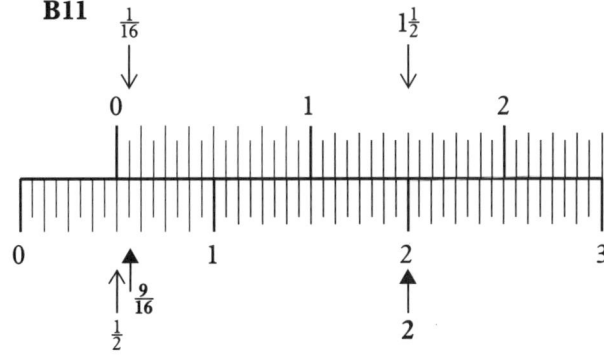

(a) $1\frac{1}{2} + \frac{1}{2} = 2$ (b) $\frac{1}{2} + \frac{1}{16} = \frac{9}{16}$

You could also have used your rulers the 'other way round'.

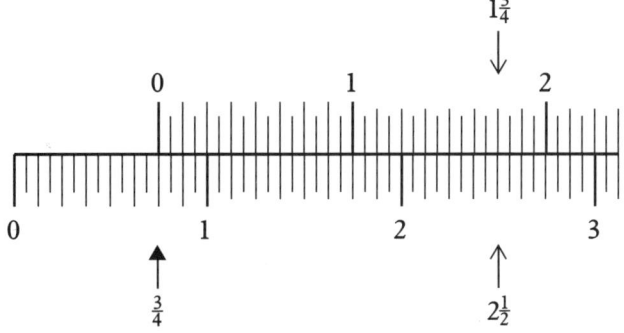

To subtract $1\frac{3}{4}$ from $2\frac{1}{2}$, place $1\frac{3}{4}$ on the top ruler against $2\frac{1}{2}$ on the bottom ruler. Read the answer on the bottom ruler against 0 on the top ruler.

$2\frac{1}{2} - 1\frac{3}{4} = \frac{3}{4}$

C1 (a) $4 + 3 + 2 = 9$ and $2\frac{3}{4} + 3 + 3\frac{1}{4} = 9$.
So the magic number is 9.

(b)

$2\frac{1}{4}$	$3\frac{1}{2}$	$3\frac{1}{4}$
4	3	2
$2\frac{3}{4}$	$2\frac{1}{2}$	$3\frac{3}{4}$

(c) If each number is doubled, the magic number is doubled. It is 18.

(d) If you subtract $\frac{1}{4}$ from each number, the magic number decreases by $3 \times \frac{1}{4} = \frac{3}{4}$. It is $8\frac{1}{4}$.

C2 Morag is right.
▲ Your own check and explanation.

C3 (a) True
▲ (b) True
(c) True
(d) False; your own example.

Challenge

Your own investigation.
Is $\frac{4}{3}$ a fraction?
What about $\frac{3}{3}$?
(See also the answer to **D4**.)

D1 (a) The line joining the fractions and point (0,0) is a straight line.

(b)

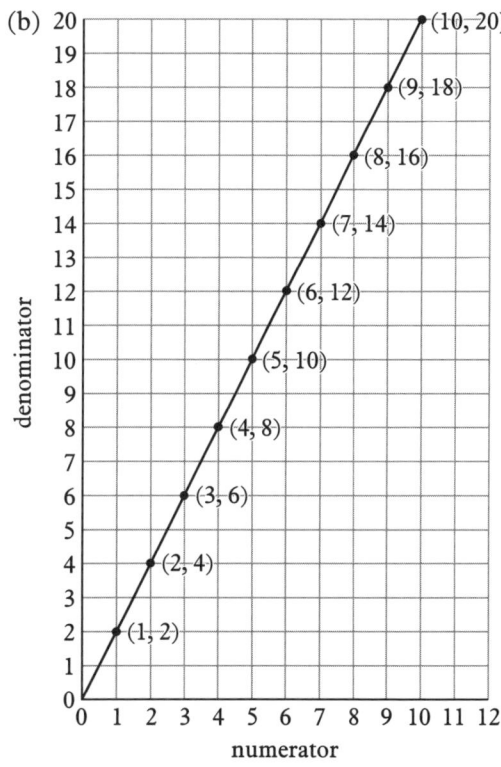

Other fractions with a value of a half are:
$\frac{3}{6} \quad \frac{5}{10} \quad \frac{6}{12} \quad \frac{7}{14} \quad \frac{8}{16} \quad \frac{9}{18} \quad \frac{10}{20}$

D2 The line joining the larger fraction to the origin is less steep.

D3

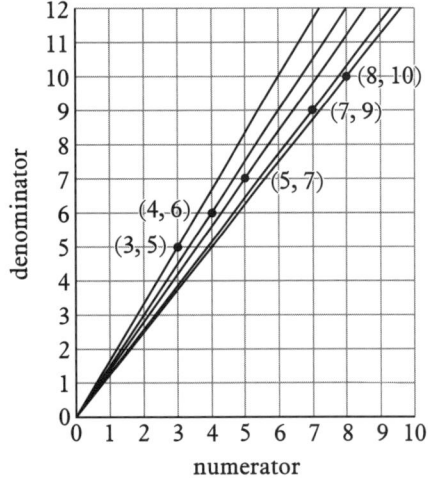

The fractions in order, smallest first, are:
$\frac{3}{5}$ $\frac{4}{6}$ $\frac{5}{7}$ $\frac{7}{9}$ $\frac{8}{10}$

D4 Your own check.

For fractions *less than* one, it is true that adding one to both the numerator and the denominator gives a larger fraction.

What happens to fractions, like $\frac{2}{2}$, which are equal to one? What happens to fractions which are larger than one?

D5

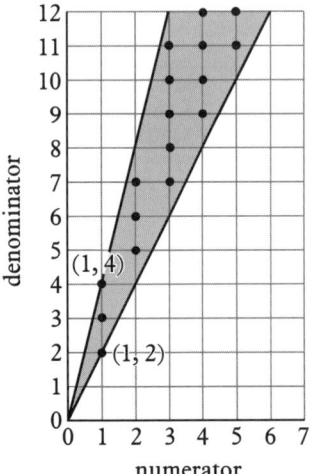

Any fraction represented by a point in the shaded area has a value between $\frac{1}{4}$ and $\frac{1}{2}$. Some of these fractions are:
$\frac{1}{3}$ $\frac{2}{5}$ $\frac{2}{6}$ $\frac{2}{7}$ $\frac{3}{7}$ $\frac{3}{8}$ $\frac{3}{9}$ $\frac{4}{9}$ $\frac{3}{10}$ $\frac{4}{10}$ $\frac{3}{11}$ $\frac{4}{11}$ $\frac{5}{11}$
$\frac{4}{12}$ $\frac{5}{12}$

Challenges

(1) Here are six to get you started:
$\frac{3}{4}$ $\frac{5}{7}$ $\frac{7}{9}$ $\frac{8}{11}$ $\frac{10}{13}$ $\frac{11}{14}$
Don't just write a list of fractions. Explain how you found them.

(2) $\frac{3}{8} = \frac{1}{4} + \frac{1}{8}$ $\frac{3}{10} = \frac{1}{5} + \frac{1}{10}$

(3) $\frac{1}{6} + \frac{1}{18} = \frac{4}{18}$ or $\frac{2}{9}$
$\frac{1}{3} + \frac{1}{5} + \frac{1}{15} = \frac{9}{15}$ or $\frac{3}{5}$

(4) Your own fractions.

Decimals 1

A1 The lizard is more than **11** cm long, but less than **12** cm long.

A2 The length of the lizard is 11 cm and **2** tenths of a cm.

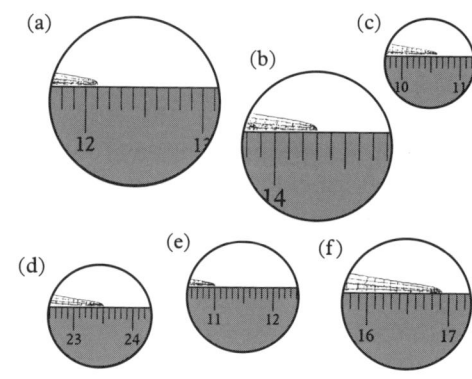

A3		cm	tenths of a cm
	(a)	12	1
	(b)	14	3
	(c)	10	6
	(d)	23	5
	(e)	11	0
	(f)	16	9

A4		Length in cm
	(a)	12·1 cm
	(b)	14·3 cm
	(c)	10·6 cm
	(d)	23·5 cm
	(e)	11·0 cm
	(f)	16·9 cm

A5

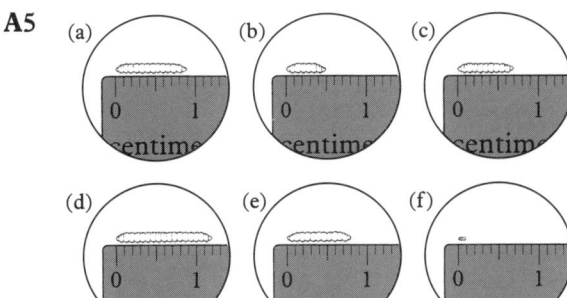

Here are the lengths of the caterpillars:
(a) 0·9 cm (b) 0·5 cm (c) 0·7 cm
(d) 1·2 cm (e) 0·8 cm (f) 0·1 cm

B1 Line *a* is **6·5** cm long.
Line *b* is **6·7** cm long.
Many people are deceived by this!

B2 *a*, *b* and *c* are all 2·6 cm long.
Can you suggest a reason why a appears to be longer?

B3

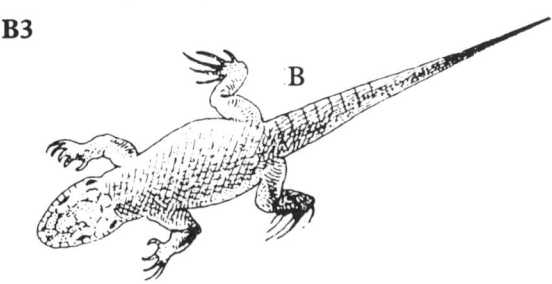

A 8·7 cm   B 7·3 cm   C 8·3 cm
D 10·2 cm   E 9·3 cm   F 8·0 cm

B4 (a) B is the shortest lizard.
▲ (b) Here is a list of the lizards in order of
        size, smallest first:
        B, F, C, A, E, D

B5 0·8 cm, 1·7 cm, 4·3 cm, 5 cm, 5·6 cm

B6 The caterpillar is 4·6 cm long.
How good was your guess?

B7 (a), (d) A is the shortest, 2·8 cm long.
(b), (c) C is the longest, 3·7 cm long.
Did you guess that A is the shortest?

B8 The real distance is 5·5 cm.

B9 Here are the real distances:
X to A 3·3 cm   X to B 7·1 cm
X to C 2·6 cm   X to D 5·7 cm

B10 and **B11** Your own guesses and work.
Are you getting better at guessing?

C1 (a) Kitten 0·4 kg   (b) Tortoise 1·8 kg
(c) Monkey 1·8 kg   (d) Alligator 3·3 kg

C2 The arrows point to these numbers:
A 1·3   B 2·2   C 3·8   D 10·7

C3 (a)
▲

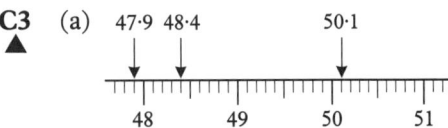

(b)

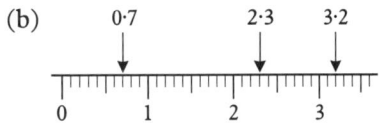

(c)

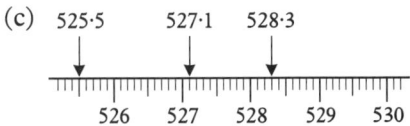

(d)

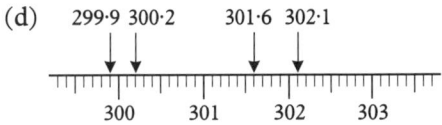

(e)

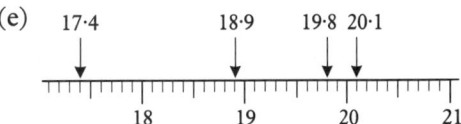

(f)

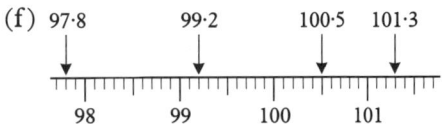

9

C4

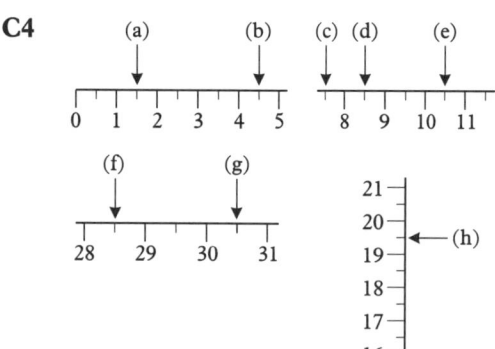

(a) 1·5 (b) 4·5 (c) 7·5 (d) 8·5
(e) 10·5 (f) 28·5 (g) 30·5 (h) 19·5

C5

		Height	Length
(a)	Kangaroo	1·8 metres	1·9 metres
(b)	Rhinoceros	1·5 metres	2·5 metres
(c)	Camel	2·0 metres	2·5 metres
(d)	Eland	2·5 metres	2·5 metres
(e)	Dragon	0·7 metres	2·9 metres

C6 (a) The giraffe is 5·8 metres high.
The man is 2·0 metres high.
The elephant is 3·6 metres high.
The bus is 4·4 metres high.
The bison is 2·1 metres high.

(b) If your guess for the length of the bus is about 10 metres, it is a good one.

C7

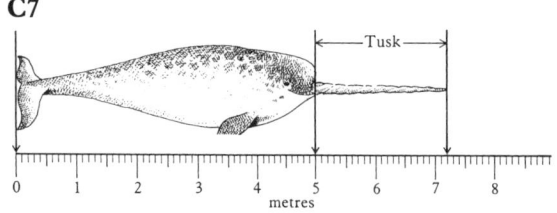

The narwhal's tusk is 2·2 metres long.

C8

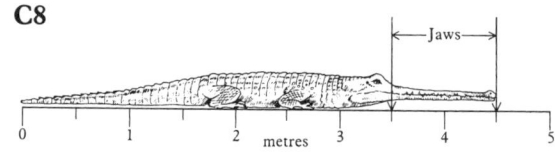

The jaws of the gavial are 1·0 metre long.

C9

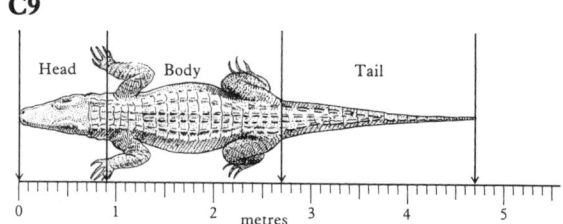

(a) Head 0·9 metre
(b) Body 1·8 metres
(c) Tail 2·0 metres

C10 (a) 2·0 metres (b) 0·6 metre
(c) 0·2 metre (d) 0·7 metre

C11 On each of the next ten days the height of the bamboo will be 9·9 m, 10·0 m, 10·1 m, 10·2 m, 10·3 m, 10·4 m, 10·5 m, 10·6 m, 10·7 m and 10·8 m respectively.

Decimals 2

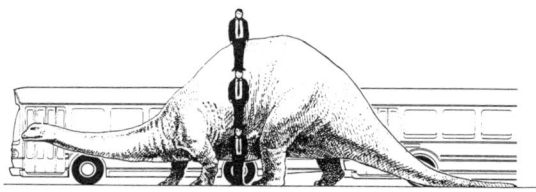

A1 The Brachiosaurus is roughly 12 to 14 metres tall.

A2 A Diplodocus is about 2½ buses long.

A3 Picture A shows the size of the Triceratops correctly.

A4 On your drawing the Polacanthus should be just less than half the length of the bus.

A5 Picture C is correct.

A6 Picture B is correct.

A7 (a) 16   (b) 100   (c) 141

A8 (a) 32   (b) 200   (c) 282

B1

Animal	Length of head and body in cm	Length of tail in cm	Name
A	7·1	7·0	House mouse
B	7·7	3·7	Common shrew
C	6·2	5·2	Harvest mouse
D	5·8	3·7	Pigmy shrew
E	9·2	9·6	Field mouse

Decimals 3

A1

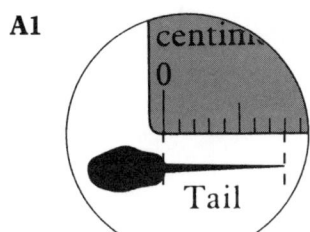

The tadpole's tail is 0·8 cm long.

A2 The total length of the tadpole is 1·3 cm.

A3 (a) $\begin{array}{r} 0\cdot7\,\text{cm} \\ +\ 1\cdot1\,\text{cm} \\ \hline 1\cdot8\,\text{cm} \end{array}$ (b) $\begin{array}{r} 1\cdot4\,\text{cm} \\ +\ 2\cdot1\,\text{cm} \\ \hline 3\cdot5\,\text{cm} \end{array}$ (c) $\begin{array}{r} 1\cdot5\,\text{cm} \\ +\ 2\cdot1\,\text{cm} \\ \hline 3\cdot6\,\text{cm} \end{array}$

  (d) $\begin{array}{r} 1\cdot6\,\text{cm} \\ +\ 1\cdot9\,\text{cm} \\ \hline 3\cdot5\,\text{cm} \end{array}$ (e) $\begin{array}{r} 1\cdot7\,\text{cm} \\ +\ 0\cdot2\,\text{cm} \\ \hline 1\cdot9\,\text{cm} \end{array}$

A4

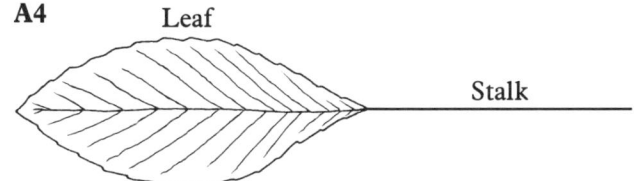

Leaf   Stalk

(a) $\begin{array}{r} 4\cdot6\,\text{cm} \\ +\ 3\cdot5\,\text{cm} \\ \hline 8\cdot1\,\text{cm} \end{array}$ (b) Your own check.

A5 (a) 5·6 cm   (b) 3·0 cm
   (c) 5·6 cm + 3·0 cm = 8·6 cm
   (d) Your own check.

A6 (a) $\begin{array}{r} 3\cdot7\,\text{cm} \\ +\ 1\cdot4\,\text{cm} \\ \hline 5\cdot1\,\text{cm} \end{array}$ (b) $\begin{array}{r} 3\cdot5\,\text{cm} \\ +\ 3\cdot0\,\text{cm} \\ \hline 6\cdot5\,\text{cm} \end{array}$ (c) $\begin{array}{r} 8\cdot0\,\text{cm} \\ +\ 3\cdot8\,\text{cm} \\ \hline 11\cdot8\,\text{cm} \end{array}$

Your own checks.

A7 (a) $\begin{array}{r} 2\cdot6\,\text{cm} \\ 0\cdot8\,\text{cm} \\ +\ 5\cdot0\,\text{cm} \\ \hline 8\cdot4\,\text{cm} \end{array}$

A8 (a) 3·7 + 2 + 0·1 = 5·8
   (b) 6·3 + 8·5 + 0·9 = 15·7
   (c) 8 + 3·6 + 1·5 = 13·1
   (d) 23·5 + 16 + 0·7 = 40·2
   (e) 0·9 + 0·8 + 3 = 4·7
   (f) 1·2 + 4·3 + 0·5 = 6
   (g) 30·4 + 16 + 0·6 = 47
   (h) 20·2 + 18·4 + 2·4 = 41

A9 (a) $\begin{array}{r} 0\cdot8\,\text{cm} \\ 4\cdot3\,\text{cm} \\ +\ 8\cdot0\,\text{cm} \\ \hline 13\cdot1\,\text{cm} \end{array}$ (b) $\begin{array}{r} 0\cdot7\,\text{cm} \\ 4\cdot6\,\text{cm} \\ +\ 7\cdot3\,\text{cm} \\ \hline 13\cdot6\,\text{cm} \end{array}$

B1

A 3·3 litres   B 2·5 litres   C 1·9 litres
D 4·0 litres

B2

There are 3·2 litres of orange in pictures E and F together.

B3 1·5 litres + 1·7 litres = 3·2 litres, which is the same answer as in **B2**.

B4

In G and H together there are 4·1 litres.

B5 (a) 2·3 + 1·8 = 4·1   (b) 3·7 + 1·7 = 5·4
(c) 0·8 + 1·4 = 2·2   (d) 3·5 + 2·9 = 6·4
(e) 4·6 + 8·6 = 13·2   (f) 3·8 + 1·2 = 5·0

C1 2·6 litres   **C2** 3·5 litres   **C3** 2·9 litres

C4 (a) 1·3 (b) 2·4 (c) 3·1 (d) 8·6
        − 0·8   − 0·6   − 0·9   − 0·8
        ____   ____   ____   ____
        0·5   1·8   2·2   7·8

(e) 2·3 (f) 3·4 (g) 4·8 (h) 5·1
     − 1·6   − 1·7   − 2·7   − 1·2
     ____   ____   ____   ____
     0·7   1·7   2·1   3·9

C5 (a) 6·2 − 1·5 = 4·7   (b) 3·7 − 1·8 = 1·9
(c) 4·2 − 2·6 = 1·6   (d) 8·4 − 3·8 = 4·6
(e) 7·0 − 1·3 = 5·7   (f) 8·0 − 2·4 = 5·6

C6 (a)   8·0   **C7**   7·0   **C8**   4·9
            − 2·6   − 3·1   − 2·0
            ____   ____   ____
            5·4   3·9   2·9

C9 (a)   7·0 (b)   8·0 (c)   9·3 (d)   6·7
            − 0·5   − 2·8   − 4·0   − 5·0
            ____   ____   ____   ____
            6·5   5·2   5·3   1·7

C10   13·0
        − 1·8
        ____
         11·2

C11 (a)   25·0   (b)   23·6
             − 2·7   − 7·0
             ____   ____
              22·3   16·6

C12 (a) The total weight is 4·2 kg.
         (b) The kitten weighs 1·3 kg.

D1   4·5
        × 3
        ____
        13·5

D2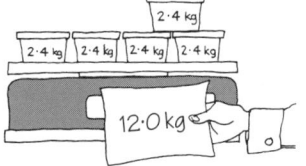

D3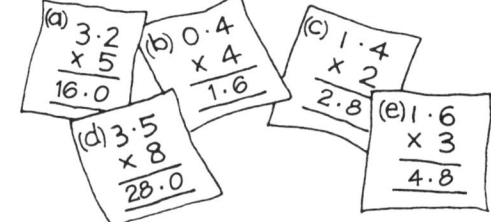

D4 (a) 25·2   (b) 2·4   (c) 13·5   (d) 20·6
(e) 57·5   (f) 27·0   (g) 22·2   (h) 47·0
(i) 1·4   (j) 11·2

D5 The total length of the five pipes is 23 metres.   (4·6 × 5 = 23)

E1 (a)   1·7   (b)   1·3   (c)   2·8
            3)5·1   4)5·2   3)8·4

E2   1·4
        ▲ 3)4·2

12

E3

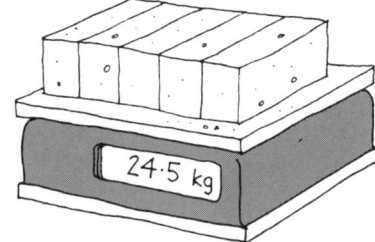

Each piece of cheese weighs 4·9 kg.

E4

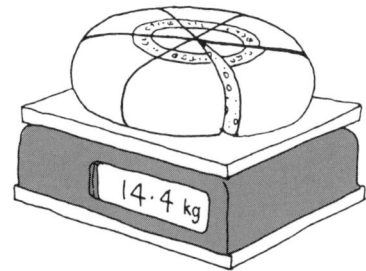

(a) Each piece weighs 2·4 kg.
(b) Each piece is $\frac{1}{6}$ of the cheese.

E5
$$\begin{array}{r} 2{\cdot}4 \\ 5\overline{)12{\cdot}0} \end{array}$$

E6 (a) $\begin{array}{r} 4{\cdot}6 \\ 5\overline{)23{\cdot}0} \end{array}$ (b) $\begin{array}{r} 4{\cdot}5 \\ 4\overline{)18{\cdot}0} \end{array}$

    (c) $\begin{array}{r} 14{\cdot}5 \\ 2\overline{)29{\cdot}0} \end{array}$ (d) $\begin{array}{r} 13{\cdot}5 \\ 4\overline{)54{\cdot}0} \end{array}$

F1 (a) $\begin{array}{r} 6{\cdot}7 \\ +\,2{\cdot}5 \\ \hline 9{\cdot}2 \end{array}$ (b) $\begin{array}{r} 48 \\ -\,17{\cdot}6 \\ \hline 30{\cdot}4 \end{array}$ (c) $\begin{array}{r} 20{\cdot}6 \\ 4\overline{)82{\cdot}4} \end{array}$

    (d) $\begin{array}{r} 28{\cdot}9 \\ -\,4 \\ \hline 24{\cdot}9 \end{array}$ (e) $\begin{array}{r} 0{\cdot}4 \\ 5\overline{)2{\cdot}0} \end{array}$ (f) $\begin{array}{r} 38{\cdot}3 \\ -\,17{\cdot}9 \\ \hline 20{\cdot}4 \end{array}$

Did you find what Paul did wrong each time?

F2 (a) 11·8   (b) 57·6   (c) 18·7   (d) 7·7

F3 (a) 11·7   (b) 98·0   (c) 10·4   (d) 21·6

F4 (a) 23·8   (b) 215·4   (c) 12·9   (d) 8·4
    (e) 3·6   (f) 19·9   (g) 8·8   (h) 1·25

Ratio and sharing 1

A1

(a) The ratio of boys to dogs is 1 to 3.
(b) The ratio of dogs to boys is 3 to 1.

A2

(a) The ratio of cars to motorbikes is 2 to 3.
(b) The ratio of motorbikes to cars is 3 to 2.

A3

(a) The ratio of windows to chimneys is
    5 to 2.
(b) The ratio of chimneys to windows is
    2 to 5.

A4 (a) The ratio of black beads to green beads
      is 1 to 2.
    (b) 24 are green.
    (c) 30 beads (20 green and 10 black)

A5 (a) The ratio of white to green is 4 to 1.
    (b) There are 80 whites.

A6 (a) The ratio of black beads to white beads is 3 to 1.
    (b) 27 black beads
    (c) 6 white beads

A7 There are 8 boys.

B1 (a) There are 2 green and 3 white cubes in each group.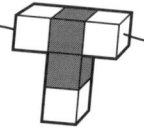
    (b) John has used 25 cubes. 10 are green and 15 are white.

B2 The ratio of white cubes to green cubes is 3 to 2.

B3 The ratio of green cubes to white cubes is 2 to 3.

B4 (a) Ann can make six groups.
       $(30 \div 5 = 6)$
    (b) She will need to paint (i) 18 white and (ii) 12 green.
    (c) Your own check using multilink cubes.
    (d) No. Ann could not use up 28 cubes to make a necklace with groups of 5 because 5 does not divide exactly into 28.
       She could make a necklace using groups of different sizes. Which sizes are possible?

B5 Jon is not right. Adding a black cube does not change the numbers of white and green cubes. The ratio of white to green is still 3 to 2.

B6

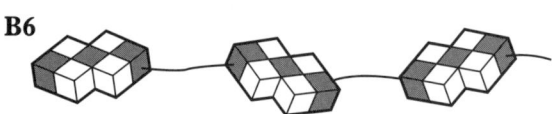

    (a) The ratio of green to white cubes is 3 to 4.
    (b) Jon will need 15 green and 20 white cubes.

B7 Your own designs.

B8 The pattern can be formed by repeating a unit like this:

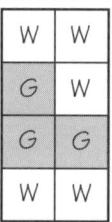

So the ratio of white to green squares is 5 : 3.

Draw a different unit which could be repeated to produce the same pattern.

C1 (a)   A   True
        B   True
        C   False
        D   False
        E   False
        F   True

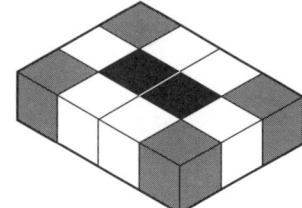

    (b) Your own true sentences.
        These could include:
        The ratio of white to green cubes is 6 : 4 or 3 : 2.
        The ratio of green to white cubes is 2 : 3.
        The ratio of black to white cubes is 2 : 6 or 1 : 3.
        The ratio of white to black cubes is 3 : 1.
        There are three times as many white cubes as black cubes.
        There are one-third as many black as white cubes.
        One-sixth of the cubes are black.
        One-third of the cubes are green.
        There are half as many black as green cubes.

C2 Your own true sentences. You should be able to write at least ten.
Notice that the ratio of black to green cubes is 1 : 1.

C3 Your own shapes.
▲ Here are two possible shapes:

The ratio of green to black cubes is 6:3 or 2:1.

The ratio of green to black cubes is 8:4 or 2:1.
The ratio of green to black cubes is always equal to 2:1.

C4 (a)

 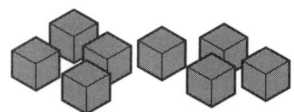

No, there are now twice as many green cubes as white cubes.
(b) Yes   (4 white and 12 green)
(c) No   (6 white and 10 green)
(d) No   (0 white and 4 green)
(e) Yes   (1 white and 3 green)
(f) Yes   (6 white and 18 green)
(g) No   (5 white and 9 green)
Multiplying the number of cubes in each pile by 4, 5, 6, ... will leave three times as many green cubes as white cubes.

C5 Emma's drink is the same strength as
▲ Ela's.
4:6 is the same ratio as 2:3.

C6

C and D have the same strength.
(1:2 = 3:6)

B, E and F have the same strength.
(1:1 = 2:2 = 3:3)

C7
▲

To answer these questions you will have to assume that the bottles and jugs all hold the same amount of liquid!
(a) False. The drinks all have the same strength: two bottles of squash to every three jugs of water.
(b) True, if you assume that the volume of squash in each bottle is the same as the volume of water in each jug.
(c) True, if you make the same assumption as in (b). The ratio 4:6 is the same as 2:3.
(d) True; 6:9 = 4:6 = 2:3.
(e) False; 2:3 is the same as 4:6.
(f) False; the strength is the same because 10:15 = 2:3.
(g) False; 20:30 = 2:3 so the strength remains the same.

C8 A and C   (1:5 = 2:10 = 10:50)

15

C9 The strength of tea depends on the volume of water as well as on the number of tea bags. The following recipes make tea of the same strength:

Number of tea bags	Number of cupfuls of water	Number of people each receiving 1 cup of tea
3	4	4
6	8	8
9	12	12
12	16	16
⋮	⋮	⋮

C10 ▲ (a) No, there is more wool than cotton. Viyella has 55 parts of wool to 45 parts of cotton.
(b) $55:45 = 11:9$

C11 The ratio of cotton to wool is $8:2$ or $4:1$.

C12 (a) The ratio of wool to polyester is $65:35$ or $13:7$.
(b) The ratio of cotton to polyester is $7:3$.

C13 ▲

	Ratio of blue to yellow
Very dark green	5:2
Dark green	2:1
Middle green	1:1
Light green	1:2
Very light green	2:5

(a) Middle green (Blue to yellow $= 2:2$ $= 1:1$)
(b) Light green (Blue to yellow $= 5:10$ $= 1:2$)
(c) Light green (Blue to yellow $= 5:10$ $= 1:2$)
(d) Dark green (Blue to yellow $= 12:6$ $= 2:1$)

The answers to (b) and (d) assume that the tins are all the same size.

C14 Very light green (Blue to yellow $= 4:10$ $= 2:5$)

Challenges

(1) (a) The mixture for sinks and basins is stronger. (30 ml of fluid to 5 litres of water is equal to 6 ml of fluid to 1 litre of water.)
(b) The ratio of fluid to water in the stain remover is $1:1000$.

(2) (a) Mirrors (b) $16:1$
(c) $6:16 = 3:8$

D1 ▲ (a) 1000 mm (b) 1 metre
(c) 15 mm or 1·5 cm

Challenge

Your own rough estimate.
Explain how you made your estimate. State any information you used and show any calculation you needed to do. An estimate of between $1:150\,000\,000$ and $1:200\,000\,000$ is reasonable.
If you are stuck, ask your teacher.

D2 ▲ $1:100$ would be a sensible scale. $1:100$ is 1 cm to 1 metre, so this scale is easy to use.
Estimate the size of your classroom and check whether a plan drawn to this scale will fit on your paper.
Could you use a larger scale?

D3 A one to ten thousand scale map will take up less paper than a one to five thousand scale map of the same village but, because it is smaller, it will show *less* detail. So Astrid is wrong about the detail on the map.

D4 (a) The $5\frac{1}{2}$ inch gauge will give the largest model.

   (b) $76 \times 27\,\text{mm} = 2052\,\text{mm}$
   The diameter of the driving wheels on the full-size engine is about 2050 mm or 2·05 m.
   The reliability of this answer depends on the accuracy of Amanda's measurement.

   (c) $1520 \div 16 = 95$
   The diameter on a $3\frac{1}{2}$ inch gauge model would be 95 mm.

Your own accounts of your investigations.

(1) I will not remain twice my daughter's age. When my daughter's age doubles, the ratio of my age to my daughter's age will be 3:2.

(2) The ratio of each rectangle's width to the length of its diagonal is 3:5.

(3) No, unless Kelly started with the same numbers of green and white cubes.

Options

The mixing of cement, garden composts and the fuel for two-stroke engines are examples of ratios used in everyday life. The scales on maps vary a lot, but these are possible ratios:

(a) countries 1:3 000 000
(b) motorways 1:250 000
(c) towns 1:5000, 1:10 000, 1:25 000

Ratio 2

A1
(a)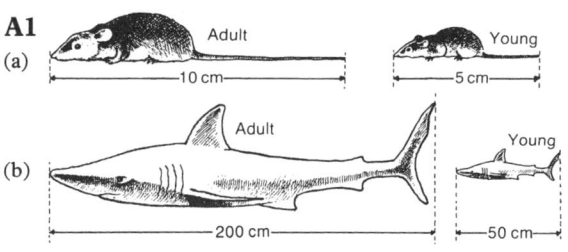

(b)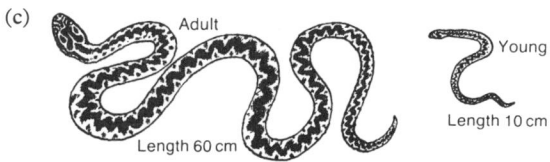

(c)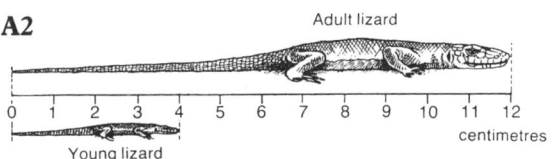

(a) Adult's length = 2 × young one's length
(b) Adult's length = 4 × young one's length
(c) Adult's length = 6 × young one's length

A2

Adult lizard

Young lizard

(a) Adult lizard's length $= 3 \times$ young lizard's length
(b) Young lizard's length $= \frac{1}{3}$ of adult lizard's length

A3

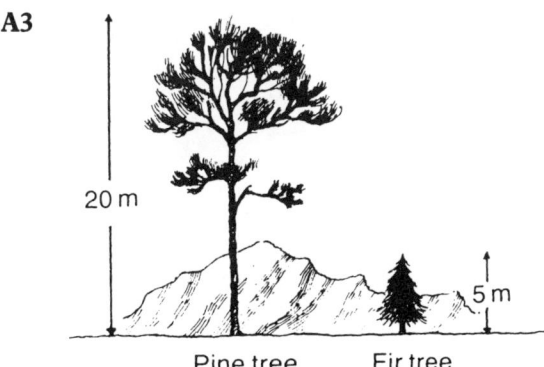

20 m

5 m

Pine tree   Fir tree

Fir tree's height $= \frac{1}{4}$ of pine's height.

A4

(a) Baby hare's weight = $\frac{1}{3}$ of adult's weight
(b) Adult hare's weight = **3** × baby's weight

(c) Adult badger's weight
    = **6** × baby's weight
(d) Baby badger's weight
    = $\frac{1}{6}$ of adult's weight

(e) Adult bear's weight = 10 × cub's weight
(f) Cub's weight = $\frac{1}{10}$ of adult's weight

A5 Kitten's weight
    = $\frac{1}{4}$ × cat's weight

A6 (a) $\frac{1}{2}$ (b) 2   (c) 3   (d) $\frac{1}{3}$
▲
A7 (a) 3   (b) $\frac{1}{3}$ (c) 2   (d) $\frac{1}{2}$
     (e) $\frac{1}{8}$ (f) $\frac{1}{3}$

A8 (a) H   (b) B

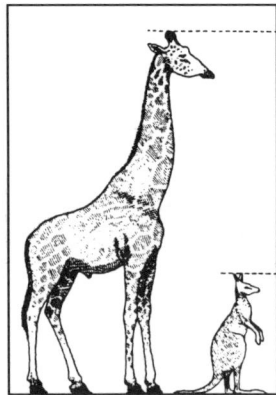

B1 (a) The kangaroo's height is 3 cm.
    (b) The giraffe's height is 9 cm.

B2 (a) The kangaroo's height is 4 cm.
    (b) The giraffe's height is 12 cm.
    (c) Yes, the kangaroo's height is still
        $\frac{1}{3}$ × the giraffe's height.

B3 The kangaroo is 5 cm tall.

B4 When the giraffe is 24 cm tall, the
    kangaroo is 8 cm tall.

B5 When the kangaroo is 10 cm tall, the
    giraffe is 30 cm tall.

B6 If the real giraffe is 4·5 metres tall, the real
▲ kangaroo is 1·5 metres tall.

B7

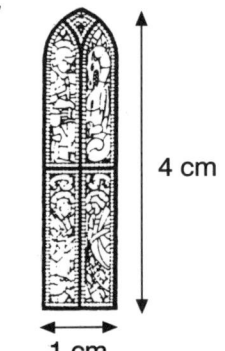

4 cm

1 cm

Width of window = $\frac{1}{4}$ × its height

B8 (a) The width of picture B is 3 cm.
    (b) Its height is 12 cm.
    (c) The width is $\frac{1}{4}$ of the height.

B9 (a) Picture C: width $2\frac{1}{2}$ cm or 2·5 cm
    (b) Picture C: height 10 cm
    (c) Yes, the width is still $\frac{1}{4}$ of the height.

B10 (a) Picture D is 4 cm wide.
(b) The whole window is 16 cm high.

B11 The real window is 2·1 metres wide.

C1
▲

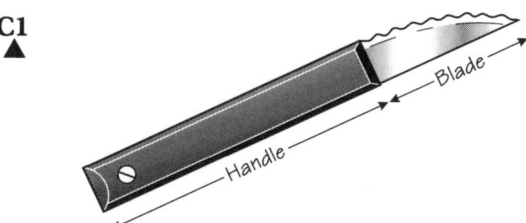

(a) Length of handle = **2** × length of blade
(b) B, D, G and H show the same knife.

C2 (a) Height = **3** × width
(b) B and F show the same doorway.

C3

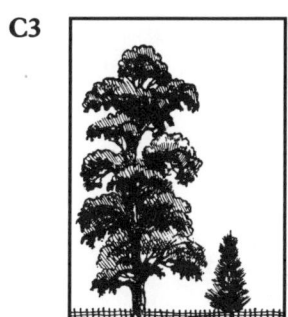

(a) Height of tall tree
 = **3** × height of short tree
(b) The real tree is 21 metres tall.

C4 (a) Young eel's length = $\frac{1}{5}$ × adult's length
(b) The length of the young eel is 0·3 m.

C5

(a) Length of bus = **3** × height of bus
(b) The real bus is 9 metres long.

C6 The handle of the broom is 120 cm or
1·2 m long.

C7 C and F could be pictures of the window.

C8 B, C, F and G could all be plans of the
swimming pool.

Fractions and decimals 1

A1

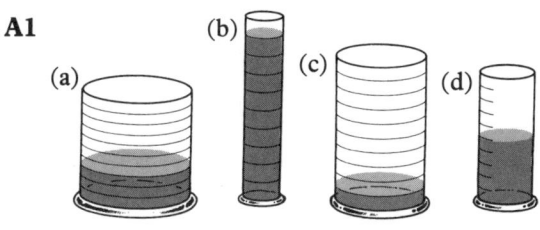

(a) is 0·3 full. (b) is 0·9 full.
(c) is 0·1 full. (d) is 0·5 full.

A2 (a) Gold-coloured 0·4 (b) Black 0·1
(c) White 0·3 (d) Grey 0·2

A3

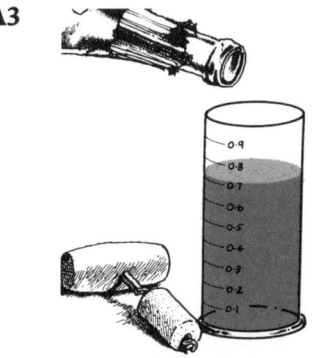

The wine bottle holds between **0·7** and **0·8**
litre.

A4 (a) 15 hundredths = 0·15
(b) 39 hundredths = 0·39
(c) 98 hundredths = 0·98
(d) 40 hundredths = 0·40

A5

	First mark	Second mark	Half-way mark
(a)	0·6	0·7	0·65
(b)	0·3	0·4	0·35
(c)	0·1	0·2	0·15
(d)	0·8	0·9	0·85
(e)	0·9	1·0	0·95

A6 There are 4 tenths and 3 extra hundredths
in 0·43.

A7 In 0·28 there are 2 tenths and 8 extra
hundredths.

A8 The nearest tenth mark to 0·47 is 0·5.

A9

	Decimal	Nearest tenth mark
(a)	0·53	0·5
(b)	0·92	0·9
(c)	0·18	0·2

A10 There are 2 tenths and 6 extra hundredths in 0·26.

B1
(a) 3 hundredths   0·03
(b) 7 hundredths   0·07
(c) 1 hundredth   0·01
(d) 18 hundredths   0·18
(e) 4 hundredths   0·04
(f) 9 hundredths   0·09

B2 0·05 is half-way between 0 and 0·1.

B3 0·17, 0·16, 0·15, **0·14, 0·13, 0·12, 0·11, 0·10, 0·09, 0·08, 0·07, 0·06, 0·05, 0·04, 0·03, 0·02, 0·01, 0·00**

B4 0·32 is higher.

B5 0·8 is higher.

B6 0·3 is highest.

B7 ▲ (a) 0·87   (b) 0·79   (c) 0·54   (d) 0·33
     (e) 0·13   (f) 0·08   (g) 0·02

B8 There is 0·47 litre of water in the jar.

B9 (a) 0·04 litre

     (b) 0·93 litre

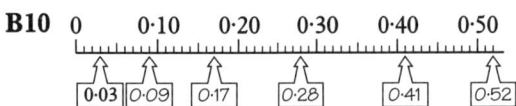

B10

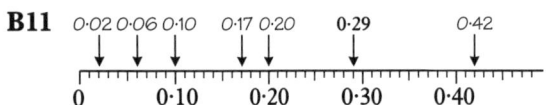

B11

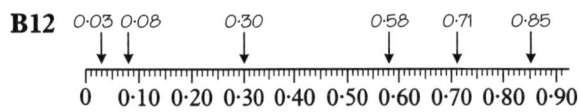

B12

B13 0·03,   0·08,   0·30,   0·58,   0·71,   0·85

B14 0·05,   0·17,   0·32,   0·46,   0·50

C1 The alligator's tail is 0·40 m long.

C2 The length of its head is (a) 15 cm, (b) 0·15 m.

C3 The body is 0·30 m long.

C4

	Name	Height
(a)	Mrs Burt	1·63 m
(b)	Dave	1·45 m
(c)	Jill	1·29 m

C5 Sue is 1·02 m tall.

C6
(a) 1 m 5 cm   **1·05 m**
(b) 1 m 50 cm   **1·50 m**
(c) 1 m 7 cm   **1·07 m**
(d) 1 m 2 cm   **1·02 m**
(e) 1 m 70 cm   **1·70 m**
(f) 1 m 20 cm   **1·20 m**
(g) 1 m 43 cm   **1·43 m**
(h) 1 m 1 cm   **1·01 m**
(i) 1 m 9 cm   **1·09 m**

C7

The length of the snake is 1·38 m.

C8

	Height in centimetres	Height in metres and centimetres	Height in metres
Mr Cray	183 cm	1 m 83 cm	1·83 m
Mrs Cray	**168 cm**	1 m 68 cm	**1·68 m**
Karen	106 cm	**1 m 6 cm**	1·06 m
Ian	**103 cm**	1 m 3 cm	**1·03 m**

C9 Here are the names of the Burts and Crays in decreasing order of height: Mr Cray, Mr Burt, Mrs Cray, Mrs Burt, Dave, Jill, Alan, Karen, Ian, Sue.

D1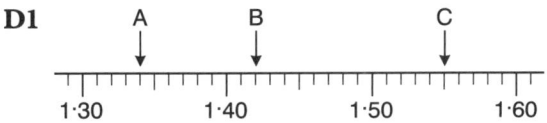

B points to 1·42 and C points to 1·55.

D2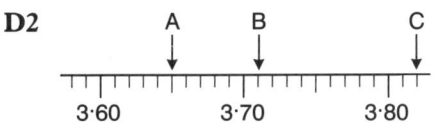

A 3·65   B 3·71   C 3·82

D3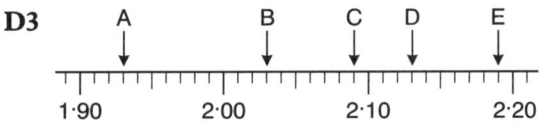

A 1·93   B 2·03   C 2·09   D 2·13   E 2·19

D4 A 4·05   B 4·15   C 4·49   D 4·83

D5 A 6·09   B 6·35   C 6·72   D 6·90

E1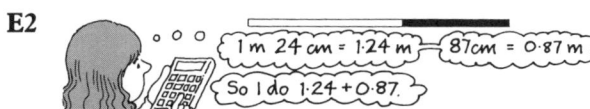

$1·38\,m + 0·94\,m = 2·32\,m$

E2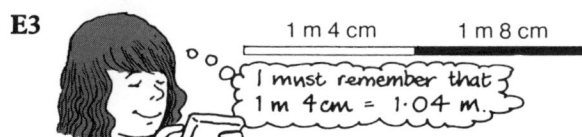

$1·24\,m + 0·87\,m = 2·11\,m$

E3

$1\,m\,4\,cm + 1\,m\,8\,cm = 1·04\,m + 1·08\,m$
$= 2·12\,m$

E4 (a) $1·68\,m + 1·03\,m = 2·71\,m$
 (b) $1\,m\,12\,cm + 85\,cm = 1·12\,m + 0·85\,m$
   $= 1·97\,m$
 (c) $1\,m\,5\,cm + 1\,m\,62\,cm = 1·05\,m + 1·62\,m$
   $= 2·67\,m$

E5 (a) 0·87 is between 0·8 and 0·9.
 (b) 0·64 is between 0·6 and 0·7.

E6

	First number	Second number	Half-way number
(a)	0·2	0·3	0·25
(b)	0·5	0·6	0·55
(c)	0·9	1·0	0·95
(d)	0	0·1	0·05

E7 In order of size, smallest first, the numbers are   0·08,   0·3,   0·8.

E8 These are the numbers in order of size, smallest first:
  0·07,   0·09,   0·7,   0·85

E9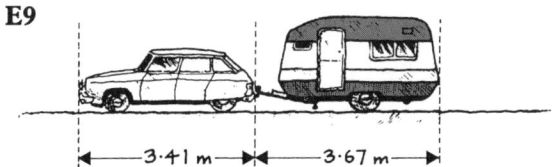

(a) The total length of the car and caravan is 7·08 m.
(b) The cost of taking them on the ferry is £7·30.

E10

(a) The total height of the lorry is 4·29 m.
(b) Yes, it can go under a bridge 4·5 m high.

E11 No, the Fizi Cola lorry is 4·40 m high.

E12 (a)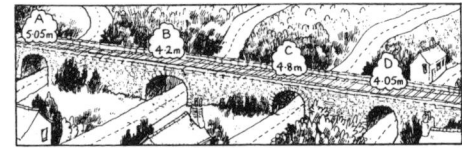

Bridge D is the lowest.

(b)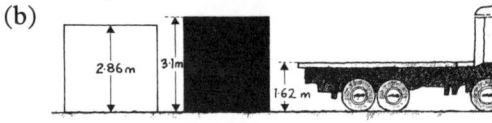

The total height will be 4·48 m.

(c) With the white box on it, the lorry can go under bridges A and C.

(d) With the black box on it, the lorry can still go under bridges A and C.
(3·1 + 1·62 = 4·72)

(e)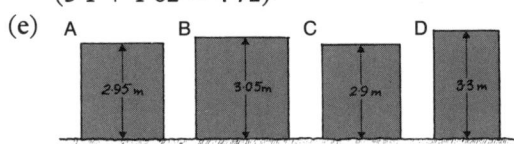

The lorry can go through bridge C with boxes A, B or C.
(4·8 − 1·62 = 3·18)

Fractions and decimals 2

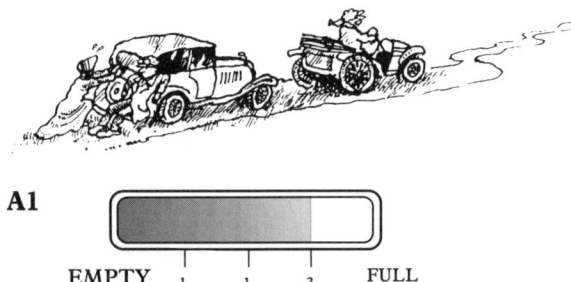

A1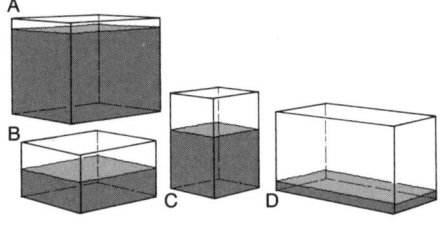

EMPTY $\frac{1}{4}$ $\frac{1}{2}$ $\frac{3}{4}$ FULL

A2 There are about 3 gallons in the tank.

A3 Mary's tank holds 7 gallons.

A4 She puts in about 7 gallons.

A5 The Vauxhall Vector has the biggest tank.

A6 The Ford Angle holds $4\frac{1}{2}$ gallons when it is half full.

A7 $\frac{4}{4}$ means 4 quarters full; in other words completely full.

A8 The grey car is a Formula 2.

A9 The blue car is a Vauxhall Vector.

A10 The black car is a Morris Minus.

B1 (a) John has walked 17 miles.
(b) He has another 26 miles to go.

B2 Rose has covered about $\frac{1}{5}$ of the walk.

B3 *John has done between $\frac{3}{8}$ and $\frac{4}{8}$ of the walk.*

B4 Rose is more than half-way.

B5 Rose gets £6·80.

C1 The pool is 30 m long.

C2 Mary has swum about 21 m.

C3 Mary has swum $\frac{7}{10}$ of the length of the pool.

C4 (a) Jane has swum $\frac{5}{10}$ of the length.
(b) $\frac{1}{2}$ is a simpler way to write $\frac{5}{10}$.

C5 *Dave is 4 tenths of a length ahead of Jane.*

C6 (a) $\frac{3}{10}$ of a length is about **9 m**.
(b) $\frac{7}{10}$ of the pool remains to be swum.

D1 (a) *Between $\frac{3}{10}$ and $\frac{4}{10}$ of the tank is full.*
(b) 28 litres   (c) $\frac{3}{10}$
(d) Roughly $\frac{8}{10}$ (e) 12 litres

D2

	Fraction full	Tank
(a)	$\frac{1}{10}$	D
(b)	$\frac{9}{10}$	A
(c)	$\frac{6}{10}$	C
(d)	$\frac{5}{10}$	B

D3 (a) $\frac{1}{10}$ of 60 = 6   (b) $\frac{8}{10}$ of 60 = 48
(c) $\frac{3}{10}$ of 60 = 18

D4 (a) 49   (b) 36   (c) 12   (d) 42
(e) 4   (f) 56

D5

	Number	$\frac{1}{10}$ of number
(a)	80	8
(b)	50	5
(c)	20	2
(d)	100	10
(e)	200	20
(f)	400	40

D6

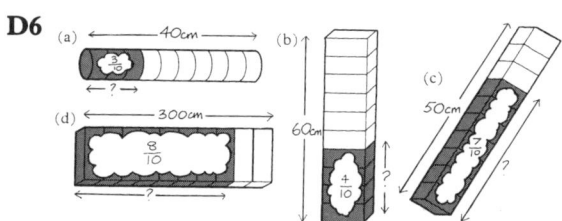

(a) **?** = 12 cm   (b) **?** = 24 cm
(c) **?** = 35 cm   (d) **?** = 240 cm

D7   (a) 24   (b) 14   (c) 27   (d) 120

D8   (a) $\frac{7}{10}$ of the bar
(b) $\frac{7}{10}$ of 120 grams = 7 lots of 12 grams
                            = 84 grams

D9   (a) 45 grams   (b) 84 grams

E1   (a) $0·9 = \frac{9}{10}$ (b) $0·3 = \frac{3}{10}$
(c) $0·5 = \frac{5}{10}$

E2   (a) $\frac{2}{10} = 0·2$ (b) $\frac{8}{10} = 0·8$
(c) $\frac{6}{10} = 0·6$

E3   (a) John has run **40 m**.
(b) Mary has run **0·7** of the track.
(c) Shaheena still has **0·1** of the track to run.
(d) Josie is **60 m** behind Shaheena.
(e) 0·6 of 100 metres is **60 m**.

E4

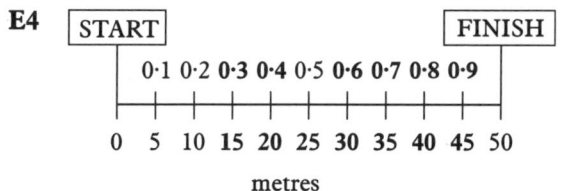

(a) 0·4 of 50 m = 20 m
(b) 0·7 of 50 m = 35 m
(c) 45 m is **0·9** of 50 m.

E5   (a) **Sadia** is 0·6 up the chimney.
(b) She is **48 m** from the ground.

E6   (a) Gary is **64 m** from the ground.
(b) He has climbed **0·8** of its height.

E7   (a) 32 m   (b) 56 m

E8   *0·9 of 80 m =* **72 m**

E9   (a) $0·1 \text{ of } 60 = \frac{1}{10} \text{ of } 60 = 6$
(b) $0·1 \text{ of } 30 = \frac{1}{10} \text{ of } 30 = 3$
(c) $0·1 \text{ of } 100 = \frac{1}{10} \text{ of } 100 = 10$

E10   ▲

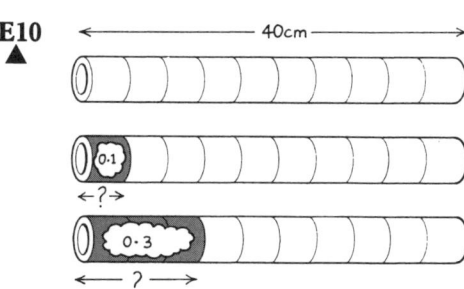

(a) *0·1 of 40 cm =* **4 cm**
(b) *0·3 of 40 cm =* **12 cm**

E11   (a) 0·4 of 50 cm = 20 cm
(b) 0·4 of 80 cm = 32 cm

E12

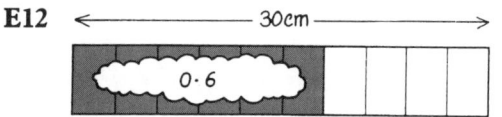

The coloured part is 18 cm long.

E13   (a) 14   (b) 30   (c) 32
(d) 60   (e) 63   (f) 120

F1   *60 per cent means the same as* $\frac{60}{100}$.

F2

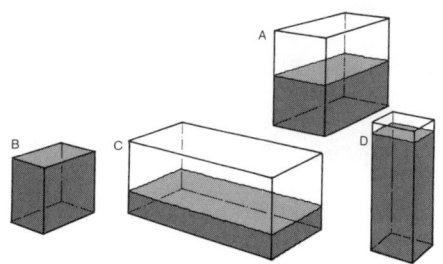

	Per cent full	Tank
(a)	100	B
(b)	50	A
(c)	30	C
(d)	90	D

23

F3 (a) 20 litres   (b) 10 litres
    (c) The tank is **25** per cent full when it has 5 litres in it.
    (d) 75 per cent.

F4 (a) 40 litres   (b) 20 litres   (c) 60 litres

F5 (a) 10 litres   (b) 15 litres

F6 (a) Jason is **8 m** from the ground.
    (b) Alison has climbed **75%** of the total height.
    (c) Rita is **6 m** above the ground
    (d) 22 m is 55% of the total height.
    (e) Shailesh still has **45%** of the height to climb.
    (f) 5% of 40 m is **2 m**.
    (g) 34 metres is **85%** of 40 metres.

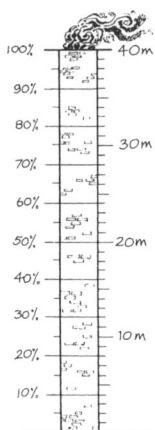

Fractions and decimals: extension 1

A1

The wingspan of the butterfly is roughly 7·3 cm.

A2

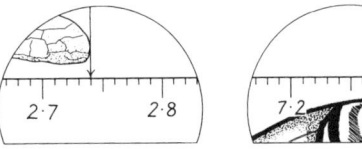

(a) The wingspan of the bee is 2·74 cm.
(b) The butterfly's wingspan is 7·27 cm.

A3

2·45 cm   2·54 cm

(a) The sheep louse is 0·09 cm long.
(b) Its leg is covering the 2·5 mark.

A4

Needle

0 0·1 0·2 0·3 0·4 0·5

0·01 0·02 0·03 0·04 0·05

(a) The feather louse is 0·25 cm long (to the nearest hundredth of a centimetre).
(b) Its head is 0·043 cm long (to the nearest thousandth of a centimetre).

A5 (a) 2·134   (b) 2·144   (c) 2·147
    (d) 0·043   (e) 0·055   (f) 1·760
    (g) 1·772   (h) 3·240   (i) 3·250

A6

2·4   2·5
(a)

2·0   3·0
(b)

2·40   2·41
(c)

On the scales the arrows point to:
(a) 2·47   (b) 2·7   (c) 2·407

A7	First number	Second number	Half-way number
(a)	3·6	3·7	3·65
(b)	4·23	4·24	4·235
(c)	7·13	7·14	7·135
(d)	1·70	1·71	1·705
(e)	8·9	9	8·95
(f)	6·4	6·41	6·405
(g)	4·99	5	4·995
(h)	0·7	0·71	0·705

B1

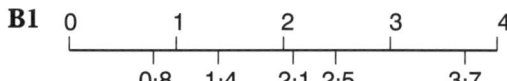

This scale is slightly reduced. Your scale should be the same size as the one on the inside back cover.

B2

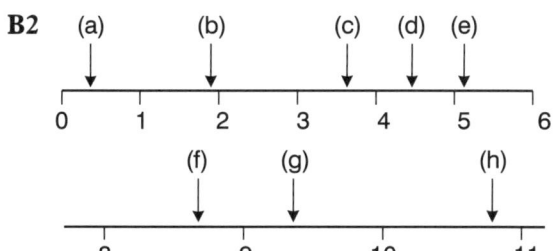

These are the estimates to the nearest tenth:

(a) 0·3 or 0·4   (b) 1·9   (c) 3·6 or 3·7
(d) 4·5   (e) 5·1   (f) 8·7
(g) 9·4   (h) 10·8

B3 (a) 3·3   (b) 4·1   (c) 4·7
(d) 5·3   (e) 5·7

B4 The length of the needle is 4·73 cm.

B5 (a) 2·72   (b) 2·96   (c) 3·05
(d) 3·19   (e) 3·31   (f) 18·87
(g) 18·99   (h) 19·04   (i) 19·15 or 19·16

B6 (a) 7·276   (b) 7·281   (c) 7·303
(d) 7·316   (e) 5·985   (f) 5·997
(g) 6·008   (h) 6·025   (i) 6·031

C1 ▲

	Number	To nearest tenth
(a)	3·26	3·3
(b)	3·21	3·2
(c)	0·58	0·6
(d)	9·44	9·4
(e)	7·07	7·1
(f)	0·09	0·1
(g)	5·96	6·0
(h)	29·98	30·0

C2

	Number	To nearest tenth
(a)	4·85	4·9
(b)	3·05	3·1
(c)	2·95	3·0
(d)	0·05	0·1

C3

	Number	To nearest hundredth
(a)	3·629	3·63
(b)	0·251	0·25
(c)	0·083	0·08
(d)	0·048	0·05
(e)	7·002	7·00
(f)	6·994	6·99
(g)	6·997	7·00
(h)	2·449	2·45

C4

	Number	To nearest hundredth
(a)	6·305	6·31
(b)	7·275	7·28
(c)	2·935	2·94
(d)	8·195	8·20
(e)	0·065	0·07
(f)	1·995	2·00
(g)	7·095	7·10
(h)	2·345	2·35

C5

	Number	To 1 decimal place
(a)	4·378	4·4
(b)	5·203	5·2
(c)	6·492	6·5
(d)	0·054	0·1
(e)	30·309	30·3
(f)	6·447	6·4
(g)	2·009	2·0
(h)	8·982	9·0

C6 (a) 4·55   (b) 1·61   (c) 28·35
(d) 4046·86   (e) 1016·05   (f) 0·30
(g) 0·45

D1 ▲

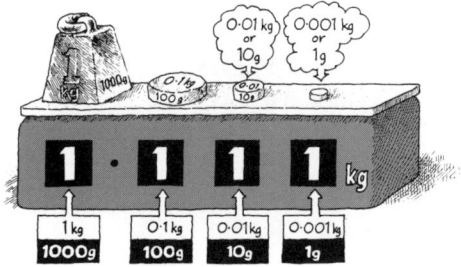

(a) 4 g = 0·004 kg
(b) 25 g = 0·025 kg
(c) 570 g = 0·570 kg
(d) 2493 g = 2·493 kg

25

D2 and **D3**

	Weight	To nearest hundredth of a kg	To nearest tenth of a kg
(a)	1·573 kg	1·57 kg	1·6 kg
(b)	0·068 kg	0·07 kg	0·1 kg
(c)	7·439 kg	7·44 kg	7·4 kg
(d)	623 g	0·62 kg	0·6 kg
(e)	48 g	0·05 kg	0·0 kg
(f)	7 g	0·01 kg	0·0 kg

D4 (a) 10·1 kg (b) 1·001 kg (c) 1100·07 kg
      (d) 0·014 kg (e) 0·17 kg

		Distance in m	Distance in km
E1		1200	1·2
E2	(a)	1700	1·7
	(b)	800	0·8
	(c)	2400	2·4
	(d)	7100	7·1

		Distance in km	Distance in m
E3	(a)	2·3	2300
	(b)	0·6	600
	(c)	0·65	650
	(d)	3·7	3700
E4	(a)	4·93	4930
	(b)	7·062	7062
	(c)	15·5	15 500
E5	(a)	0·13	130
	(b)	0·04	40
	(c)	0·007	7

E6 (a) 723 m = 0·723 km
      (b) 54 m = 0·054 km
      (c) 13 250 m = 13·250 km
      (d) 610 m = 0·61 km
      (e) 90 m = 0·09 km
      (f) 400 m = 0·4 km

E7 ▲ (a) 3·62 km = 3620 m
      (b) 28 m = 0·028 km
      (c) 0·903 km = 903 m
      (d) 41·6 km = 41 600 m
      (e) 73 400 m = 73·4 km
      (f) 2 m = 0·002 km
      (g) 4·70 km = 4700 m
      (h) 21 050 m = 21·05 km

E8 (a) 1663 mm = 1·663 m
      (b) 904 mm = 0·904 m
      (c) 900 mm = 0·9 m
      (d) 46 mm = 0·046 m

E9 (a) 6 m = 6000 mm
      (b) 0·65 m = 650 mm
      (c) 1·2 m = 1200 mm
      (d) 0·382 m = 382 mm

E10 (a) 1·62 km = 1620 m
       (b) 75 mm = 0·075 m
       (c) 0·02 m = 20 mm
       (d) 4370 m = 4·37 km

E11
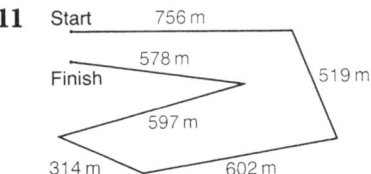

The total length of the course, to the nearest 0·1 km, is 3·4 km.

E12 (a) The distance from Slough House Farm to South Hanningfield Church is 2·81 km, to 2 decimal places.
       (b) 2·81 km is 2810 m.

E13 (a) 3·06 km   (b) 3060 m

E14 (a) 2·59 km   (b) 2590 m

E15 The distance across the reservoir is about 3050 m.

E16 (a) Your estimate.
       (b) The distance around Manningfield Reservoir is about 12 km.

Fractions and decimals: extension 2

A1 (a) $314·65 \times 10 = 3146·5$
      (b) $8·079 \times 10 = 80·79$
      (c) $51·907 \times 10 = 519·07$
      (d) $0·086 \times 10 = 0·86$

A2

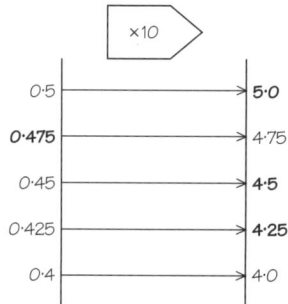

A3
(a) 18	(b) 180	(c) 1·55
(d) 15·5	(e) 11·5	(f) 115
(g) 0·68	(h) 68	(i) 0·35
(j) 3·5	(k) 1	(l) 10

A4

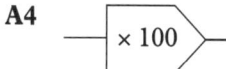

A5
(a) 259·1	(b) 3736	(c) 4003·25
(d) 8·7	(e) 11 320	

A6 Multiplying by 1000 moves figures **three** places to the left.
(a) 6304·6	(b) 237·1	(c) 12 360
(d) 87	(e) 81 324	

A7
(a) 0·863	(b) 2·502	(c) 0·0586

A8
(a) 0·1295	(b) 0·067	(c) 0·0018

A9
(a) $4·71 \div 100 = 0·0471$
(b) $0·083 \div 100 = 0·00083$
(c) $100 \times 0·0624 = 6·24$
(d) $58·6 \div 1000 = 0·0586$
(e) $1000 \times 0·403 = 403$
(f) $1·2063 \times 100 = 120·63$
(g) $1000 \times 0·052 = 52$
(h) $100 \times 7·003 = 700·3$
(i) $6·058 \div 100 = 0·06058$

B1
(a) 0·1 of 6 m is 0·6 m.
(b) 0·2 of 6 m is 1·2 m.
(c) 0·3 of 6 m is 1·8 m.
(d) 0·4 of 6 m is 2·4 m.
(e) 0·5 of 6 m is 3·0 m.
(f) 0·6 of 6 m is 3·6 m.

B2
(a) 0·1 of 4 m = 0·4 m
(b) 0·2 of 4 m = 0·8 m
(c) 0·6 of 4 m = 2·4 m
(d) 0·9 of 4 m = 3·6 m

B3
(a) 0·3 of 4 m = 1·2 m
(b) 0·7 of 4 m = 2·8 m

B4
(a) 0·6 of 36 litres = 21·6 litres
(b) 0·9 of 36 litres = 32·4 litres

B5
(a) 0·3 of 41 kg = 12·3 kg
(b) 0·8 of 24 metres = 19·2 m

B6 ▲ 0·6 of £54 = £32·40

B7
(a)
$$\begin{array}{r} 23 \\ \times\ 3 \\ \hline 69 \end{array}$$
so 0·3 of 23 = 6·9

(b)
$$\begin{array}{r} 19 \\ \times\ 4 \\ \hline 76 \end{array}$$
so 0·4 of 19 = 7·6

(c)
$$\begin{array}{r} 40 \\ \times\ 8 \\ \hline 320 \end{array}$$
so 0·8 of 40 = 32

B8

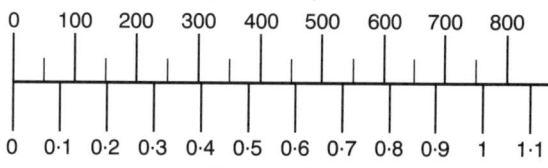

(a) Mach 0·3 is about 230 m.p.h.
(b) Mach 0·7 is about 530 m.p.h.
(c) Mach 0·8 is about 610 m.p.h.

B9
(a)
$$\begin{array}{r} 760 \\ \times\ 3 \\ \hline 2280 \end{array}$$
so 0·3 of 760 m.p.h. = 228 m.p.h.

(b)
$$\begin{array}{r} 760 \\ \times\ 7 \\ \hline 5320 \end{array}$$
so 0·7 of 760 m.p.h. = 532 m.p.h.

(c)
$$\begin{array}{r} 760 \\ \times\ 8 \\ \hline 6080 \end{array}$$
so 0·8 of 760 m.p.h. = 608 m.p.h.

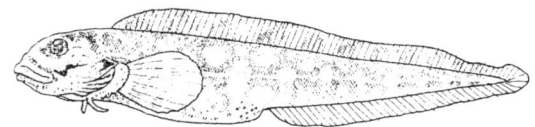

B10 $2.4 \times 12 = (2 \times 12) + (0.4 \times 12)$
$$= 24 + 4.8$$
$$= 28.8$$

B11 (a) $3.1 \times 28 = (3 \times 28) + (0.1 \times 28)$
$$= 84 + 2.8$$
$$= 86.8$$
(b) $4.2 \times 33 = (4 \times 33) + (0.2 \times 33)$
$$= 132 + 6.6$$
$$= 138.6$$
(c) 117.0 (d) 157.7 (e) 59.2 (f) 99.0

B12

	1m	2m	3m	4m	5m	6m	7m	8m	9m
Heavy duty	33	66	99	132	165	198	231	264	297
Extra thick	28	56	84	112	140	168	196	224	252
Thick	25	50	75	100	125	150	175	200	225
Medium	20	40	60	80	100	120	140	160	180
Thin	16	32	48	64	80	96	112	128	144

(a) 3 m of thin wire cost 48p, so 30 m will cost **£4·80**.
(b) 7 m of thick wire cost 175p, so 0·7 m will cost **18p**, to the nearest penny.

B13

		Length	Cost
(a)	Thin	28·4 m	£4·54
(b)	Extra thick	39·2 m	£10·98
(c)	Heavy duty	40·8 m	£13·46
(d)	Thick	160·3 m	£40·08

C1 Length of Dwarf's stick $= 0.8 \times 45$ cm
$$= 36 \text{ cm}$$

C2 Length of Dwarf's body $= 0.8 \times 15$ cm
$$= 12 \text{ cm}$$

C3 Height of Dwarf's fuse $= 0.8 \times 32$ cm
$$= 25.6 \text{ cm}$$

C4 $a = 18$, $b = 12$, $c = 6$, $d = 4.8$

C5 $a = 1.8$, $b = 1.08$, $c = 0.48$

C6 House A   (a) 4·5 cm   (b) 6·0 cm
           (c) 1·5 cm   (d) 0·8 cm
       House B   (a) 3·15 cm   (b) 4·2 cm
           (c) 1·05 cm   (d) 0·56 cm

D1 (a) $3 \times 8 = 24$
  $\div 10$ ... $\div 10$
  $0.3 \times 8 = 2.4$

(b) $4 \times 5 = 20$
  $\div 10$ ... $\div 10$
  $4 \times 0.5 = 2.0$

(c) $6 \times 0.3 = 1.8$
  $\div 10$ ... $\div 10$
  $0.6 \times 0.3 = 0.18$

(d) $0.2 \times 3 = 0.6$
  $\div 10$ ... $\div 10$
  $0.2 \times 0.3 = 0.06$

D2 $8 \times 7 = 56$
  $\div 10$ ... $\div 10$
  $0.8 \times 7 = 5.6$
  $\times 10$ ... $\times 10$
  $0.8 \times 70 = 56$

D3 $4 \times 2 = 8$
  $\div 10$ ... $\div 10$
  $4 \times 0.2 = 0.8$
  $\div 10$ ... $\div 10$
  $0.4 \times 0.2 = 0.08$

You could divide the 4 by 10 first.

D4

$$3 \quad \times \quad 6 \quad = \quad 18$$

(÷10)   (÷10)

$$3 \quad \times \quad 0.6 \quad = \quad 1.8$$

(÷10)   (÷10)

$$3 \quad \times \quad 0.06 \quad = \quad 0.18$$

(×10)   (×10)

$$30 \quad \times \quad 0.06 \quad = \quad 1.8$$

D5 (a) $0.3 \times 800 = 240$ (b) $0.07 \times 30 = 2.1$
(c) $600 \times 0.09 = 54$ (d) $400 \times 0.7 = 280$
(e) $0.8 \times 0.2 = 0.16$ (f) $0.3 \times 0.3 = 0.09$
(g) $60 \times 0.05 = 3$

E1 ▲ (a) $\frac{8}{0.2} = \frac{80}{2} = 40$ (b) $\frac{24}{0.6} = \frac{240}{6} = 40$

(c) $\frac{3}{0.1} = \frac{30}{1} = 30$ (d) $\frac{48}{1.2} = \frac{480}{12} = 40$

E2 (a) $\frac{4.8}{0.06} = \frac{48}{0.6} = \frac{480}{6} = 80$

(b) 700 (c) 7 (d) 7000 (e) 3
(f) 0.5 (g) 600 (h) 0.03

E3 400 pieces $(320 \div 0.8 = 400)$

E4 90 bottles $(27 \div 0.3 = 90)$

E5 350 sheets $(3.5 \div 0.01 = 350)$

E6 150 buttons $(3 \div 0.02 = 150)$

E7 The kangaroo will take **500** jumps to cover 3 km. $(3\,\text{km} = 3000\,\text{m}; \quad 3000 \div 6 = 500)$

E8 The snail will take **23** minutes.
$(4.6\,\text{m} = 460\,\text{cm}; \quad 460 \div 20 = 23)$

E9 (a) If no toffee is wasted, **1500** bars can be made. $(60\,000 \div 40 = 1500)$
(b) 5% of 60 kg = 3 kg = 3000 g;
so 95% of 60 kg = 60 000 g – 3000 g
$= 57\,000\,\text{g}$
With 5% wastage, $57\,000 \div 40 = \mathbf{1425}$
bars of toffee are made.

F1 (a) The height of A is 1·2 cm.
Your own measurement.
(b) To work out the height of A, you divide by 3.
(c) The length of A is 7·5 cm ÷ 3 = 2·5 cm.

F2 (a) To undo × 0·4, you **divide by 0·4.**
(b) The height of C is 5·2 cm ÷ 0·4 = 13 cm.
(c) The width of D is 2 cm, so the width of C is 2 cm ÷ 0·4 = 5 cm.

F3 (a) The length of the picture is 8·4 cm.
The length of the postcard is given by
8·4 cm ÷ 0·6 = 14 cm.
(b) The height of the picture is 4·8 cm and that of the postcard 8 cm.

F4 The real photo is:
(a) 19 cm tall (b) 14 cm wide
(c) From corner to corner it measures about 23·6 cm.

Fractions and decimals: extension 3

A1 (a)

0   1
5 parts
(b) $\frac{5}{15} = \frac{1}{3}$
You may have found other fractions.

A2 (a)

0   1
6 parts
(b) Another fraction equal to $\frac{1}{3}$ is $\frac{6}{18}$.

A3

$$\frac{1}{4} = \frac{2}{8} = \frac{3}{12} = \frac{4}{16} = \frac{5}{20} = \frac{6}{24}$$

A4

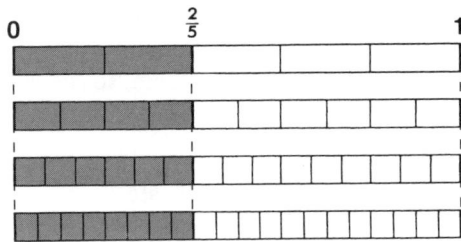

$\frac{2}{5} = \frac{4}{10} = \frac{6}{15} = \frac{8}{20}$

A5 (a)
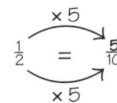
$\frac{1}{2} = \frac{5}{10}$ (×5)

(b) $\frac{2}{3} = \frac{8}{12}$ (×4)

(c) $\frac{12}{20} = \frac{3}{5}$ (÷4)

(d) $\frac{10}{12} = \frac{5}{6}$

(e) $\frac{5}{6} = \frac{25}{30}$ (f) $\frac{20}{30} = \frac{2}{3}$

(g) $\frac{18}{24} = \frac{3}{4}$ (h) $\frac{4}{7} = \frac{28}{49}$

(i) $\frac{4}{5} = \frac{36}{45}$

A6 (a) $\frac{9}{12} = \frac{3}{4}$ (b) $\frac{18}{30} = \frac{3}{5}$ (c) $\frac{40}{60} = \frac{2}{3}$

(d) $\frac{400}{600} = \frac{2}{3}$ (e) $\frac{27}{48} = \frac{9}{16}$ (f) $\frac{15}{27} = \frac{5}{9}$

A7

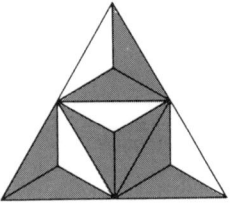

(a) The triangle is divided into 12 equal parts.
(b) 8 parts are coloured.
(c) $\frac{8}{12} = \frac{2}{3}$ of the triangle is coloured.

B1 (a) $\frac{3}{5} = \frac{12}{20}$ (b) $\frac{3}{5}$ is less than $\frac{13}{20}$.

B2 $\frac{5}{6} = \frac{20}{24}$ so $\frac{5}{6}$ is greater than $\frac{19}{24}$.

B3 $\frac{5}{8} = \frac{35}{56}$ so $\frac{5}{8}$ is less than $\frac{37}{56}$.

B4 $\frac{8}{9} = \frac{32}{36}$ so $\frac{31}{36}$ is less than $\frac{8}{9}$.

B5 $\frac{3}{7}$ is less than $\frac{4}{9}$.

B6 (a) $\frac{2}{3}$ is greater than $\frac{3}{5}$.

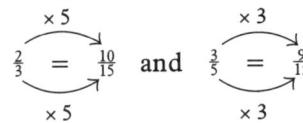

$\frac{2}{3} = \frac{10}{15}$ (×5) and $\frac{3}{5} = \frac{9}{15}$ (×3)

(b) $\frac{3}{5}$ is greater than $\frac{7}{12}$.
($\frac{3}{5} = \frac{36}{60}$ and $\frac{7}{12} = \frac{35}{60}$)

(c) $\frac{5}{7}$ is greater than $\frac{7}{10}$.
(d) $\frac{13}{15}$ is greater than $\frac{5}{6}$.

B7 (a)
$\frac{3}{20} = \frac{15}{100} = 0.15$ (×5)

(b) $\frac{9}{20} = \frac{45}{100} = 0.45$ (c) $\frac{17}{20} = \frac{85}{100} = 0.85$

B8 (a) $\frac{1}{4} = \frac{25}{100} = 0.25$ (b) $\frac{13}{50} = \frac{26}{100} = 0.26$

(c) $\frac{7}{25} = \frac{28}{100} = 0.28$

B9 (a) $\frac{7}{10} = \frac{70}{100} = 0.7$ (b) $\frac{4}{5} = \frac{80}{100} = 0.8$

(c) $\frac{19}{20} = \frac{95}{100} = 0.95$ (d) $\frac{11}{20} = \frac{55}{100} = 0.55$

(e) $\frac{3}{25} = \frac{12}{100} = 0.12$

B10 (a) $\frac{16}{25} = 0.64$ (b) $\frac{13}{20} = 0.65$

(c) $\frac{13}{20}$ is greater.

B11 (a) $\frac{3}{4} = 0.75$ (b) $\frac{18}{25} = 0.72$

(c) $\frac{3}{4}$ is greater.

C1

Each person gets $\frac{2}{3}$ of a cake.

C2 $\frac{3}{8}$ kg

C3 $\frac{4}{7}$ litre

C4 (a) $4 \div 9 = \frac{4}{9}$ (b) $3 \div 7 = \frac{3}{7}$

(c) $2 \div 5 = \frac{2}{5}$ (d) $9 \div 10 = \frac{9}{10}$

(e) $6 \div 8 = \frac{6}{8} = \frac{3}{4}$

D1 (a) $\frac{3}{8} = 0.375$ (b) $\frac{1}{8} = 0.125$

(c) $\frac{1}{4} = 0.25$ (d) $\frac{3}{4} = 0.75$

(e) $\frac{3}{5} = 0.6$ (f) $\frac{4}{9} = 0.444\ldots$

D2 $\frac{2}{3} = 0.\dot{6}$

D3 $\frac{7}{9} = 0.\dot{7}$

D4 (a) $\frac{5}{6} = 0.8\dot{3}$ (b) $\frac{5}{12} = 0.41\dot{6}$

D5 (a) $\frac{4}{11} = 0.\dot{3}\dot{6}$ (b) $\frac{1}{11} = 0.\dot{0}\dot{9}$

D6 (a) $0.\dot{1}4285\dot{7}$ (b) $0.\dot{2}8571\dot{4}$
(c) $0.\dot{5}7142\dot{8}$ (d) $0.\dot{7}1428\dot{5}$
(e) $0.\dot{8}5714\dot{2}$ (f) $0.1\dot{8}$
(g) $0.58\dot{3}$ (h) $0.0\dot{7}692\dot{3}$

Look again at the decimals for the sevenths.
What do you notice?

Percentage

A1 35% of the tank is full.

A2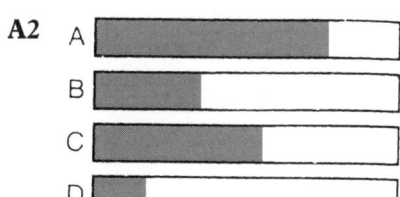

	% coloured	Strip
(a)	40%	B
(b)	80%	A
(c)	20%	D

A3 (a) **Strip Q** is 60% coloured.
(b) **Strip S** is 50% coloured.
(c) **Strip R** is 90% coloured.
(d) **Strip P** is 30% coloured.

A4

70% of the strip is coloured.

A5 The answer is on the inside back cover of the booklet.

Did you colour too much or too little? Make another tracing and, without looking at the back cover, try again. Were you closer this time?

B1 30% is nickel.

B2 (a) 50% is tin.   (b) 50% is lead.

B3 0% 10% 20% 30% 40% 50% 60% 70% 80% 90% 100%

Copper	Zinc

B4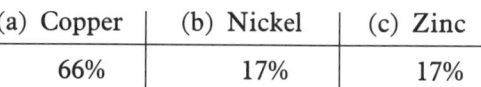

(a) Silver	(b) Copper	(c) Zinc
50%	35%	15%

Check: 50% + 35% + 15% = 100%

B5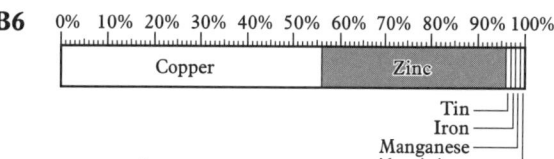

(a) Copper	(b) Nickel	(c) Zinc
66%	17%	17%

Check: 66% + 17% + 17% = 100%

B6 0% 10% 20% 30% 40% 50% 60% 70% 80% 90% 100%

Copper	Zinc	

Tin ———
Iron ———
Manganese ———
Aluminium ———

(a) Copper 56%   (b) Zinc 40%
(c) Tin 1%   (d) Iron 1%
(e) Manganese 1%   (f) Aluminium 1%

B7 36% is nickel.   (100% − 64% = 36%)

B8 8% is nickel.   (100% − 74% − 18% = 8%)

C1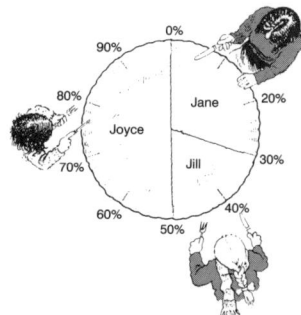

(a) 30% of the pie is Jane's.
(b) Jill's share of the pie is 20%.
(c) Joyce has 50% of the pie.

C2

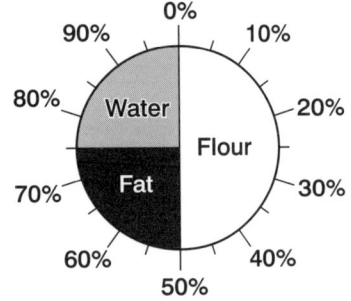

(a) Flour	(b) Fat	(c) Water
50%	25%	25%

C3

(a) Flour	(b) Butter	(c) Water	(d) Salt
41%	41%	17%	1%

C4 ▲

(a) Flour	(b) Suet	(c) Water	(d) Salt
48%	24%	26%	2%

C5 (a)   Your own title.

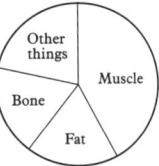

(b)   Your own title.

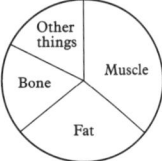

C6 (a)   Your own title.

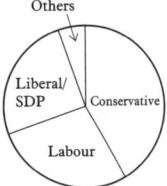

(b)   Your own title.

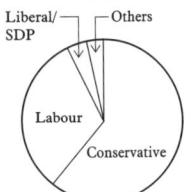

(c) The Liberal/SDP alliance has fewer MPs than you might expect.

C7 Your own pie chart.

C8 Labour 53%, Conservative 27%, SDP 20%

C9

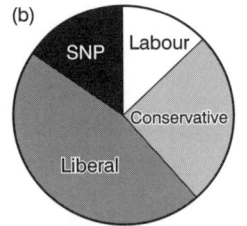

Ludlow 1983   Orkney and Shetland 1983

(a) Labour 13%, Conservative 55%, SDP 32%

(b) Labour 13%, Conservative 26%, Liberal 46%, SNP 15%

D1 (a) 80% black   (b) 30% black

D2 (a) 50% blue   (b) 30% blue

E1

	Amount	50%
(a)	£20	£10
(b)	£60	£30
(c)	84p	42p
(d)	90 kg	45 kg

E2

	Amount	25%
(a)	£80	£20
(b)	£24	£6
(c)	48p	12p
(d)	£1·20	£0·30 or 30p

E3

	Amount	75%
(a)	20p	15p
(b)	80p	60p
(c)	£16	£12
(d)	£1·60	£1·20

E4

	Amount	10%
(a)	£60	£6
(b)	£65	£6·50
(c)	£79	£7·90
(d)	£3·80	£0·38 or 38p

E5 (a) 10% of £40 is £4.
So 5% of £40 is £4 ÷ 2 = **£2**.
(b) 5% of £48 = £2·40
(c) 5% of £1·60 = £0·08 or 8p
(d) 5% of 280 kg = 14 kg

F1 (a) Using the scale, 65% of £40 = £26.
(b) 95% of £40 = £38

F2 Your own drawing of the scale in the booklet.

F3

```
0%  10% 20% 30% 40% 50% 60% 70% 80% 90% 100%
├──┼──┼──┼──┼──┼──┼──┼──┼──┼──┤
£0  £12 £24 £36 £48 £60 £72 £84 £96 £108 £120
```

F4

```
0%  10% 20% 30% 40% 50% 60% 70% 80% 90%100%
├──┼──┼──┼──┼──┼──┼──┼──┼──┼──┤
0   6   12  18  24  30  36  42  48  54  60 kg
```

G1 The exact answer is £23·56.
You should be close to this.

G2 These are the exact answers:
(a) £31·08   (b) £53·04   (c) £8·74
(d) £43·68   (b) £9·90   (f) £5·67
What was your score? 14 or over is a very good score.

H1

(a) 30%	(b) 80%	(c) 90%	(d) 10%	(e) 40%
0·3	0·8	0·9	0·1	0·4

H2 (a) 72% = 0·72   (b) 49% = 0·49
(c) 94% = 0·94   (d) 24% = 0·24
(e) 18% = 0·18   (f) 70% = 0·70

H3 (a) 3% = 0·03   (b) 1% = 0·01
(c) 8% = 0·08

H4 The cat's share is £406.

H5 (a) £195·75   (b) £123·25
(c) Check:
£406 + £195·75 + £123·25 = £725

H6 (a) £38·08   (b) £18·36   (c) £11·56
(d) Check: £38·08 + £18·36 + £11·56 = £68

H7 (a) Your own estimate. An estimate between £20 and £30 is a good one.
(b) £26·24

H8

Problem	Exact answer
43% of £29	£12.47
74% of £82	£60.68
28% of £326	£91.28
13% of £94	£12.22
58% of £65	£37.70

10 or over is a good score.

H9 (a), (b) and (c) are obviously wrong.

H10 (a) 13·78 grams of gold
(b) 5·72 grams of copper
(22% of 26 grams = 5·72 grams)

H11

(a) £1·80   (b) £2·16   (c) £4·20
(d) £1·26   (e) £22·40

H12

	Normal price	Saving (15%)
(a)	£5	£0·75 (or 75p)
(b)	£12	£1·80
(c)	£81	£12·15

H13 (a) £10 worth of 50p coins weigh 270 g.
(b) There will be 202·5 g of copper.
(75% of 270 = 202·5)

Percentage: extension

A1

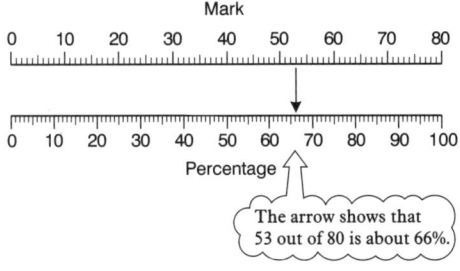

Mark
```
0  10  20  30  40  50  60  70  80
```

```
0  10  20  30  40  50  60  70  80  90  100
```
Percentage

The arrow shows that
53 out of 80 is about 66%.

  (a) 24 out of 80 = 30%

  (b) 12 out of 80 = 15%

  (c) 65 out of 80 = 81 or 82%

A2 (a) 45%   (b) 71 or 72%   (c) 15%

A3 (a) 58%   (b) 38%   (c) 91%

A4 (a) 21%   (b) 44%   (c) 65%

A5 (a) 78%   (b) 23%   (c) 47%

A6 Your own estimates. Here are the answers
to the nearest 1%.

  (a) 85%   (b) 61%   (c) 82%

  (d) 34%   (e) 41%   (f) 73%

B1

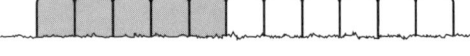

  (a) $\frac{5}{11}$ of the fence is painted.

  (b) $\frac{5}{11}$ = 0·45 to 2 decimal places

  (c) 45% of the fence is painted.

B2 (a) $\frac{14}{18}$ or $\frac{7}{9}$ of the tiles are coloured.

  (b) 78%

B3 79% of the seeds came up.

B4 (a) 77%   (b) This packet was worse.

B5 (a) Your own check.   (b) 0·23

  (c) 23%

B6 (a) 41%   (b) 5%   (c) 81%

B7 (a) The total police force was
126 400 + 12 200 = 138 600.

  (b) 91% were men. $\left(\frac{126\,400}{138\,600}\right)$

  (c) 9% were women. $\left(\frac{12\,200}{138\,600}\right)$

*Did you bother to work out the second
fraction?*

B8

	Wear glasses	Do not wear glasses
Boys	59	126
Girls	28	104

  (a) The total number of boys is 185.

  (b) 32% of the boys wear glasses.

  (c) There are 132 girls.

  (d) 21% of the girls wear glasses.

  (e) There are 317 children altogether.

  (f) 87 children wear glasses.

  (g) 27% of all the children wear glasses.

  (h) 58% of the children are boys.

B9 41% of the ring is gold.

B10 32% of the bracelet is silver.

B11 (a) No   (b) 33% is silver.

  (c) 40·8 g of alloy *should* contain 14·3 g of
silver (to 1 decimal place).

B12 (a) No, Trundles are not the most reliable.
Fewer Trundles break down than any
other make but fewer are sold.

  (b) Floodgates are the most reliable. Only
13% of Floodgates break down.
(Compare this with 29% of Trundles
breaking down.)

*Do you agree that a higher percentage of
Trundles break down than of any other
make?*

C1

Soya beans	40%	Stewing beef	17%	Rice	6·2%
Peanuts, roasted	28%	Cod and haddock	16%	Baked beans	6·0%
Cheese, Cheddar	25%	Eggs	12%	Milk	3·3%
Chicken	21%	White bread	8·3%	Potatoes	2·1%

			Amount of protein
(a)	Roasted peanuts	25 g	7 g
(b)	Baked beans	250 g	15 g
(c)	Milk	500 g	16·5 g

C2 (a) 55 g (not enough)   (b) 60 g (enough)

C3 (a) (i) 69%   (ii) 23%   (iii) 8%
    (b) (i) 45%   (ii) 49%   (iii) 6%
    (c) Neither kind of cream breaks the law.
    (d) **Single cream**

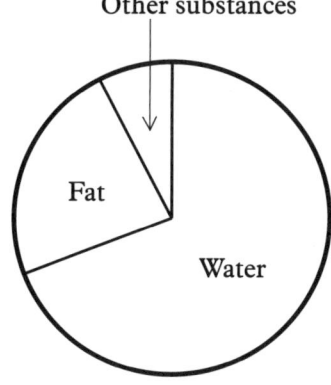

Double cream

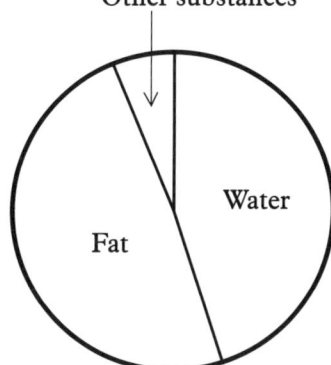

Ratio, fractions, decimals and percentages

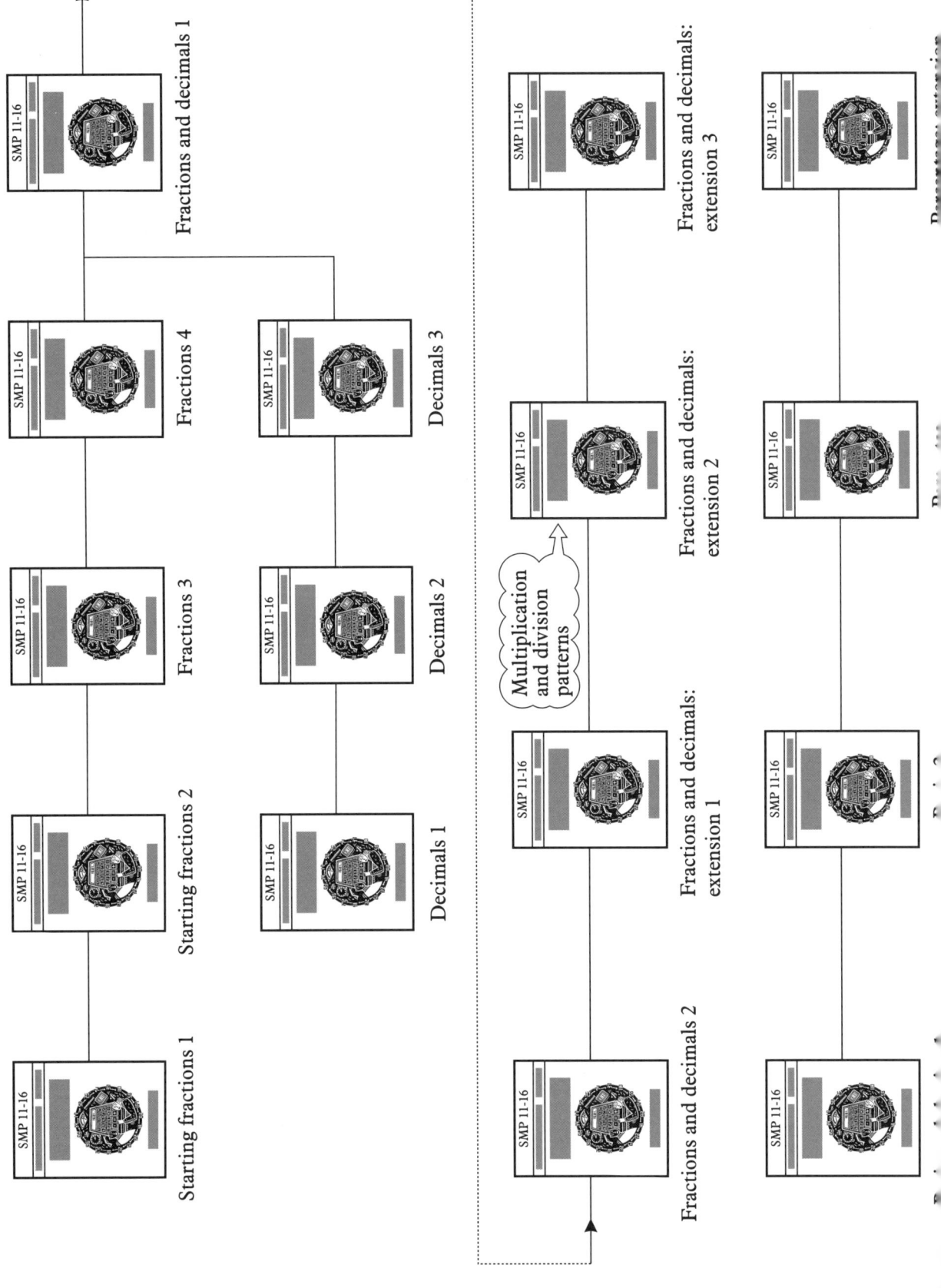

SMP 11-16	SMP 11-16	SMP 11-16	SMP 11-16	SMP 11-16
Starting fractions 1	Starting fractions 2	Fractions 3	Fractions 4	Fractions and decimals 1

SMP 11-16	SMP 11-16	SMP 11-16	SMP 11-16
Decimals 1	Decimals 2	Decimals 3	

Multiplication and division patterns

SMP 11-16	SMP 11-16	SMP 11-16	SMP 11-16
Fractions and decimals 2	Fractions and decimals: extension 1	Fractions and decimals: extension 2	Fractions and decimals: extension 3

SMP 11-16	SMP 11-16	SMP 11-16
		Percentages extension